U-Boats
in the
Bay of
Biscay

an essay in
Operations
Analysis

U-Boats
in the
Bay of
Biscay

an essay in
Operations
Analysis

Brian McCue

2008

ALIDADE
PRESS

ISBN:
Hardcover   978-1-4363-0593-8
Softcover   978-1-4363-0592-1

ALIDADE PRESS
31 Bridge Street,
Newport, Rhode Island 02840
www.alidadepress.com

Alidade Press publishes books relevant to Information Age warfare and republishes classic monographs on military affairs that are no longer in print. The Press also issues the *Information Age Warfare Quarterly*, a journal of interdisciplinary essays and analytical articles on the application of complex systems research to military operations.

Editorial Experts, Inc., Alexandria, Virginia, edited, designed, and proofread the 1990 edition of this book under contract. Shirley Kessel, Primary Sources Research, Chevy Chase, Maryland, prepared the index.

Library of Congress Cataloging-in-Publication Data
McCue, Brian.
   U-boats in the Bay of Biscay : an essay in operations analysis /
Brian McCue.
      p. cm.
   Includes bibliographical references and index.
   1. World War, 1939-1945—Campaigns—Biscay, Bay of (France and Spain) 2. Operations research—History. 3. World War, 1939-1945—Military intelligence. 4. World War, 1939-1945—Naval operations— Submarine. 5.World War, 1939-1945—Naval operations, German. I. Title.
D810.S7M37 1990
940.54'51—dc20   90-42794
                                                       CIP

First printing, September 1990, National Defense University Press
Second printing, March 2008, Alidade Press

This book was printed in the United States of America.

To order additional copies of this book, contact:
Xlibris Corporation
1-888-795-4274
www.XLIBRIS.COM
Orders@XLIBRIS.COM
45318

Contents

Tables

Figure

Dedication

To Four Great Teachers:
Miriam Crowley McCue
John Joseph Gerald McCue
Mary Keenan
John ("Doc") Walters

Preface

Neither my friend Brian McCue nor his sponsors at the National Defense University could possibly have known the great service they were rendering eighteen years ago when *U-Boats in the Bay of Biscay* was first published. Vice Admiral Jack Baldwin accurately describes the value of the quantitative analysis of operations in 1942, as well as in 1990 when the book appeared. We three with many others believed then that deadly submarine and antisubmarine campaigns are won by "outscouting the enemy"—knowing how to search for, detect, and track your target so you can sink or avoid him. But in 1990 the Soviet Union collapsed, its formidable submarine fleet decayed, and nobody seemed to care about victory by detecting, tracking, and killing a highly distributed enemy, instead of the easier analytical problem of delivering high volumes of fire efficiently against fixed targets.

Then in the past decade or so came an awareness of two things: first, the terrorist enemy has created a new scouting problem. There is much more to successful peacemaking operations than finding and quickly acting to kill, isolate, or neutralize this new form of highly distributed, mobile, and elusive enemy. Nevertheless it is fair to say solving the modern search problem is a necessary condition for success against terrorists even though it is not sufficient. Second, the power of information technology opened up new avenues to acquire, transfer, and integrate detection cues unimagined in 1942 or even in 1990: networked operations. Alidade, which is reissuing McCue's little classic, is an enthusiastic sponsor of networked thinking to achieve realistic methods and useful applications.

Not enough research has been done to check past analyses for the accuracy of their results and applicability of their conclusions. McCue does this for the U-Boat campaign in its most critical months. He has the advantage over those who wrote the earlier accounts of access to now-declassified information about the code-breaking by both sides.

McCue emphasizes that successful OA is as much art as science. B. O. Koopman in *Search and Screening* described the mathematical foundations of search theory, as he had been instructed to do when he was a young and powerful mathematician attached to Philip Morse's ASWORG. But Koopman was much more than a mathematician and some of the *art* of practical search and screening appears in his second edition written shortly before he died. Morse and George Kimball in *Methods of Operations Research* underscored the applicability of the scientific process and the utility of quantitative investigation. This was the right emphasis in the book that set this new field in motion. But they perhaps went too far in the belief that only experienced scientists could perform successful OR. Morse was skeptical that naval officers could ever be effective analysts, which of course was quite wrong. Naval officers bring their own experience to bear and have a running start over civilian analysts in knowing operational context.

In his book McCue blends the best of classical mathematical and scientific methods with the new power of the computer and due emphasis on operations analysis as high art. I believe McCue reaches all audiences better than anyone before him. On one hand he has written a subtle essay for students steeped in scientific methods of analysis who need to learn more about the artistry of practical OA. At the other end of the spectrum he has written a primer—an introduction—for those working in fields of social, policy, and military art and science. He shows these readers how mere words can be undergirded by models and analysis. Illustrating with a real campaign analysis of vital importance when the issue was very much in doubt, McCue describes the power and limitations of rigorous and clarifying descriptions of combat phenomena.

Wayne P. Hughes, Jr.
Captain, US Navy (Retired)
Dean, Graduate School of
Operational and Information Sciences
Naval Postgraduate School
Monterey California

Introduction to 2007 Edition

Search Theory was developed when operators and academics collaborated to apply mathematics to a very tangible task: to find militarily important objects in the quickest, best way possible. As Search Theory matured, two groups of professionals emerged: analysts working to solve operational search problems and academics developing new mathematical tools. For a variety of reasons, these two groups have taken different paths. Operations Research analysts continue to ply their trade with a toolbox of practical techniques among tactically minded decision makers, while the academics have gravitated to useful but abstract results in complex multivariate mathematics.

This parting of the ways was motivated largely by the fact that until recently the military search problem had not changed appreciably since the Battle of the Atlantic. In addition, some in the military community even suggested that search itself would no longer be relevant in a world where high-fidelity sensors would blanket the earth with space-age, multi-spectral intelligence satellites. This view has proven to be overly optimistic. As satellites have collected enormous amounts of data, they created a different search problem: how to discern meaningful patterns in large data sets. Similarly, low-cost surveillance drones and passive detectors have created new search problems: how best to distribute and synchronize sensors. Finally and perhaps most importantly, the global war on terrorism calls for solutions to the complex problem of locating elements of highly secretive, low technology networks in cities across the globe as well as in regions that have traditionally been of relatively low priority for military and intelligence planners.

These and other new search problems inspired a community of researchers to bring the original two professional groups together in the "New Horizons in Search Theory Workshop" series, which has convened in Newport, Rhode Island annually since 2000. Among the topics discussed at these workshops are: the intersection of classic military search strategies with new mathematical models for solving complex problems; what Search Theory might look like if it were re-invented using today's mathematical and computational resources; and Hider Theory, the obverse of Search Theory.

One of the recommendations from the early workshops was to republish one of the classic examinations of World War II search, Brian McCue's 1990 *U-Boats in the Bay of Biscay: An Essay in Operations Analysis*. Dr. McCue's study of British search operations in the Bay of Biscay is a valuable addition to Search Theory and broader operations analysis literature for several reasons. The frameworks and analyses that Dr. McCue used to evaluate the Bay of Biscay operations and to compare their relative utility to other anti-submarine and shipping-protection strategies can be adapted to help today's leaders think through operational alternatives in the global war on terror, homeland security, more traditional forms of war-fighting and the new operational strategies that are evolving as we enter the Information Age. Another useful contribution of this book is its examination of the co-evolution of technological and tactical innovation by British and German forces. As Dr. McCue points out, that while no one innovation had a lasting effect due to adaptation by the other side, the overall competitive process favored the more adaptive British.

Dr. McCue's book will also be of interest to non-specialists for its cogent analysis of a lesson that leaders need to continually re-learn: the evaluation of operational alternatives often hinges upon the measures of effectiveness. In war, as in much else, a key is correctly identifying measures of effectiveness. Lastly, *U-Boats in the Bay of Biscay* demonstrates the importance of an initial step that is too often taken for granted: solving the "problem of finding the problem". This is ironically itself a search.

This classic of search theory has been too long out of print; but republishing it is by no means a nostalgic exercise. Operations Research analysts, academics and military professionals will find useful lessons for the future in this book.

Jeff Cares
Alidade Press
Newport, Rhode Island

Foreword

One characteristic of modern war and the study of war in general is the use of quantitative tools—variously known as operations analysis, operations research, or systems analysis—to measure weapons, tactics, and strategies, as well as to evaluate actual or hypothetical battles. Introduced in World War II, these techniques have come to be associated with the military, even as they have improved and spread into nonmilitary fields. Despite such growth and widespread acceptance, operations analysis (to pick a single term) remains an imperfect representation of reality.

In this "operations analysis history," a combination of traditional history and analysis employing quantitative techniques, author Brian McCue explores the uses and limits of operations analysis. He takes as his text the 1942-44 campaign against German U-Boats in the North Atlantic, completing the analysis the pioneering WWII researchers never had a chance to finish. In the process, he validates the usefulness of their techniques even as he clarifies and identifies the limits of their analysis. In a key finding, he stresses the overwhelming importance of selecting appropriate measures of effectiveness when attempting to quantify military operations. Beyond its obvious appeal to the military operations research community, McCue's essay generates broad principles—supported by both empirical evidence and analytical modeling—of interest to national security strategists and policymakers. For example, his critical analysis of the troubles with the "top-down approach" used by current defense analysts has great currency for modern policymakers. McCue's conclusions might reasonably be extended to the measurement of other military endeavors, such as bomber operations or antimissile defense systems.

This study is not for the casual reader looking for another account of the well-documented, even romanticized, battles of the North Atlantic.

Its unique approach, bridging the gap between historical and scientific representations of reality, challenges the reader intellectually. But this study offers, in return, crisp, accessible prose and many fresh insights into the fascinating attempts of modern man to evaluate quantitatively that human enterprise most resistant of all to measurement—warfare.

J. A. BALDWIN
Vice Admiral, US Navy President,
National Defense University

Acknowledgments

This book germinated under the most pleasant of circumstances at National Defense University's Strategic Capabilities Assessment Center. For nine months the Center's Director, Bob Butterworth, gamely accepted without tangible evidence my assurances that I was doing something worthwhile involving U-boats. When I finally had something to show for my efforts he—along with Warren Dew, Ari Epstein, Mike Gilmore, Joe Goldberg, Jim Graham, Mike Harris, Fred Kiley, Martin Libicki, Steve Linke, J.J.G. McCue, Ted Postol, George Rathjens, Jack Ruina, Ron Siegel, James Tyson, and Alan Washburn—read the manuscript and suggested numerous improvements. At NDU Press, Fred Kiley, Paul Taibl, and George Maerz showed admirable restraint in their dealings with an anxious and impatient first-time author, and the NDU Library staff cheerfully aided me in my various quests for the treasures in their stacks. Their colleague at the Center for Naval Analyses, Pam Hutchins, deserves special praise for her repeated success in finding long-forgotten documents and shepherding them through the process of approval for public release and unlimited distribution.

In my U-boat project, as in so many of my undertakings, I enjoyed the benefit of Jack Nunn's gentle guidance.

The above luminaries' brilliance, however, could reach these pages only through a most refractory medium—the author, with whom responsibility for any aberrations in the final product must reside.

Brian McCue

1

Overview

The late war, more than any other, involved the interplay of new technical measures and opposing countermeasures. For example, the German U-boats had to revise their tactics and equipment when we began to use radar on our anti-submarine aircraft; and we, in turn, had to modify our tactics and radar equipment to counter their changes. In the see-saw of techniques the side which countered quickly, before the opponent had time to perfect the new tactics and weapons, had a decided advantage. Operations research, *bringing scientists in to analyze the technical import of the fluctuations between measure and countermeasure, made it possible to speed up our reaction rate in several critical cases.*
—Admiral E.J. King, 1945[1]

In the Second World War, German U-boats (Unterseeboote) operated against Allied shipping, principally convoys in the North Atlantic, in an effort to reduce shipments of basic necessities and war materiel to Great Britain. The U-boats jeopardized Great Britain's war effort so much that Winston Churchill later wrote, "The only thing that ever really frightened me during the war was the U-boat peril."[2]

The Allied response to the U-boats benefited greatly from the introduction of *operations research,* defined by its early practitioners as "a scientific method of providing executive departments with a quantitative basis for decisions regarding the operations under their control."[3] Although the last U-boats have long since surrendered, the mathematical methods used against them in the Second World War deserve further attention for three reasons.

First, the war ended before operations researchers could com-plete a full analysis of the campaign in the North Atlantic. They addressed the salient elements—the effectiveness of searches from aircraft and surface ships, the methods of attacking from these platforms once a U-boat was found, the disposition of escorts around a convoy, and so on—and formulated what today we would call a "macro-model" of U-boat circulation in the North Atlantic. But operations researchers at the time never united these analyses into a single, comprehensive study from which they or historians could draw new conclusions about the war.

Second, the wartime successes of operations research led to the postwar use of similarly structured quantitative inquiries, variously described as *operations research, systems analysis,* or *operations analysis.*[4] A sizable industry now performs such work for defense contractors, the Department of Defense, and other parts of the Federal Government. This industry can trace much of its acceptance and popularity to the success of operations research in analyzing antisubmarine warfare. Present-day heirs of the wartime operations researchers would benefit from a detailed look at their legacy; the successful application of operations research in the past is not just a selling point for today's research, it is a source of many procedural implications for today's researcher.

Third, the measure-countermeasure see-saw of which Admiral King, commander of the U.S. effort against the U-boats, wrote was not the last of its kind. Antisubmarine warfare exists to this day; in fact, the nuclear-armed ballistic missiles on some submarines enable them to pose a peril far beyond that Churchill saw in the U-boats. Other modern examples, many involving either strategic nuclear weapons, electronic warfare, or both, contain two elements—a heavy emphasis on technology and the decisiveness of any technological advantage—that so richly rewarded systematic study of the war against the U-boat. Today's policymakers can therefore learn much from generalizations drawn from a complete study of the U-boat war, including how operations research during the war differs from that practiced today.

This work aims to show that one can quantify and analyze the campaign against the U-boats not merely by applying the window dressing of "index numbers" and "coefficients based on sound military judgment" to a preconceived set of impressionistic notions, or by the mechanistic application of curve-fitting techniques, but through mathematical reasoning systematically applied to knowable physical quantities. This technique appears in the realm of strategic nuclear exchange analysis, but without verification—fortunately—from any actual strategic nuclear exchanges. In the present study the close fit between computed results and wartime outcomes shows the bounds of the former's accuracy.

Today's version of military operations research has little in common with the wartime version on whose success it trades. Part of the difference, of course, stems from the distinction between wartime and peacetime conditions and decision-making requirements. However, a clear bifurcation of intellectual content has also occurred. Whereas the work appearing in today's operations research journals shows the benefits of four decades of mathematical progress, it often takes the form of solutions looking for problems,[5] and many studies done for the military are less sophisticated mathematically. Even more disturbing, the wartime success in choosing measures of effectiveness is rarely duplicated today.

Much modern analysis aims to determine the "system requirements" for new or proposed weapon systems and to find ways to meet those requirements that cannot be defeated by enemy countermeasures. In this book I will show that, in the U-boat war, the introduction and assimilation of new hardware proceeded so slowly that an inherently defeatable device could have a useful career while the other side spent time realizing that the device had been deployed, arguing about what it was, and introducing a counterdevice. Moreover, I will argue that the "top-down" approach to weapon design, in which the design is derived from the requirements, is fundamentally inappropriate in cases involving possible enemy countermeasures.

In part because of increased U.S. academic attention to Soviet military thought,[6] the once-fuzzy distinctions among such concepts as tactics, doctrine, strategy, and grand strategy have been revisited in hope of making them more clear-cut.[7] The Bay of Biscay operations, and in fact the whole *praxis* of Admiral of Submarines Karl Dönitz, show that the modern hierarchical arrangement of military concepts may make for good scholasticism, but it does little to explain the process of prosecuting a modern war unless appropriate measures of effectiveness are applied to each level of the hierarchy.

I will also confirm the finding of wartime operations researchers that new tactics or operational policies can have as much impact as new equipment, and often more immediately. Though this finding exists in the literature, if not articulated precisely as above, only the seasonal adjustment I will introduce here prevents a skeptic from arguing that the seeming success of countermeasures really stemmed from the fact that the British introduced their innovations in the springtime, as the period of daylight lengthened, while the German countermeasures introduced some months later received an apparent boost in effectiveness from the shortening of the day.

I will also quantify other traditional notions, such as the cyclical nature of innovation, the efficacy of wolfpack tactics, and the necessity of providing adequate logistical support even for technologically dominant weapon systems such as the U-boat: the mundane considerations of repair and refueling had far more effect on the U-boats' success than did the more exciting technical developments of electronic countermeasures. I will quantify some new findings, such as the relation between the value to the British of offensive search in the Bay of Biscay and the value of decryption-based disruption of at-sea U-boat refuelings. The very quantification of these findings will highlight the importance of choosing appropriate measures of effectiveness.

The Bay of Biscay: The U-Boats' Gateway to the North Atlantic

U-boats most imperiled North Atlantic shipping in 1942 and 1943. The Battle of the Atlantic entered a new phase December 9, 1941, with Hitler's notification to U-boat Command that all restrictions on operations against U.S. shipping or in U.S. waters were lifted. (This notification came two days after Pearl Harbor and two days before the German declaration of war on the United States.[8]) U-boat operations declined greatly in late 1943, and in April 1944 Admiral King issued a report downgrading the U-boat "from a menace to a problem."[9]

The German U-boats operated from ports in occupied France, crossing the Bay of Biscay to gain access to the North Atlantic. (See figure 1.) Independent shipping could be found anywhere except in the Bay itself; convoys were best hunted in the mid-Atlantic. There, beyond the convenient reach of aircraft (and totally beyond their reach until the latter part of 1943), U-boat wolfpacks—directed from ashore by Dönitz—could seek out and attack convoys.

The U-boats' close communication with their commander was to be their undoing. As disclosed in the 1970's, the Enigma encryption device used to encipher almost all German radio transmissions during the war

Figure 1
U-Boat Operations Against Convoys
Adopted from Price, p. 128

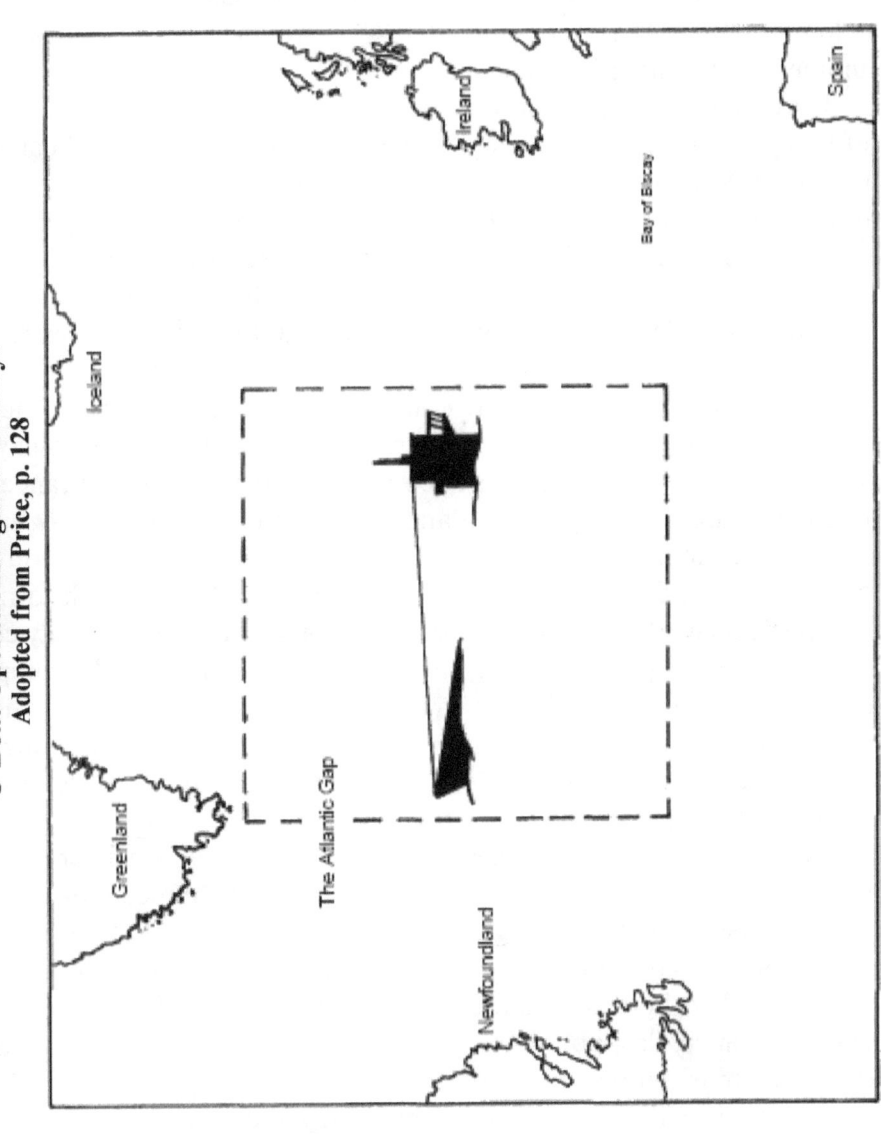

was compromised; British and American intelligence services could decipher many of these messages quickly and accurately. An upgrade of the Enigma machine made U-boat signals secure for most of 1942, but for most of 1943 the U-boat messages could be read in England as well as in Germany. (The exceptions in 1943 were three brief periods in the summer, each lasting less than a month.)

Unrefueled, U-boats could remain at sea for more than a month; at-sea refueling from a specially designed tanker submarine could add another month to the cruise. This added time at sea contributed disproportionately to mission effectiveness because it was spent in or near the operational area. Refueling at sea eliminated the need to spend 20 or more days on the trip to port and back. In Dönitz's words, the extended time at sea "advanced our Biscay bases anything from 1,000 to 2,000 miles farther westwards."[10]

The passage through the Bay could cost a U-boat more than time and fuel. From 1941 to 1944, Allied aircraft searched the Bay of Biscay for German submarines in transit between their ports in occupied France and their patrol areas in the North Atlantic. Only in the Bay of Biscay, in the area near Trinidad, in the Straits of Gibraltar, and near the Allied convoys themselves could Allied aircraft find submarines often enough to make search efforts worthwhile. (See figure 2.) Though the U-boat density could hardly be called profuse—early in the war an aircrew could complete an entire tour of duty without ever sighting a U-boat, much less sinking one[11]—the "offensive" campaign in the Bay proved fruitful enough to warrant the continued diversion of aircraft and crews from the "defensive" task of protecting convoys.

Repairs at the U-boat bases in occupied France consistently took longer than Dönitz would have liked. A shortage of trained men, as well as occasional Allied bombing, lengthened the turnaround time at these French ports. When the ports were abandoned in August 1944, a few U-boats had to be destroyed to prevent their capture by Allied forces retaking France, despite furious efforts to rig all the boats with snorkels and get them under way.[12]

As Admiral King stated, the technology the two sides used in the U-boat campaign changed constantly throughout the war. Though this book will address the whole Atlantic campaign, the study of this technological tug-of-war will focus on offensive search in the Bay from January 1942 to January 1944, in order to examine the measure-countermeasure battle

Figure 2
Bay of Biscay
From Sternhell and Thorndike, p. 143

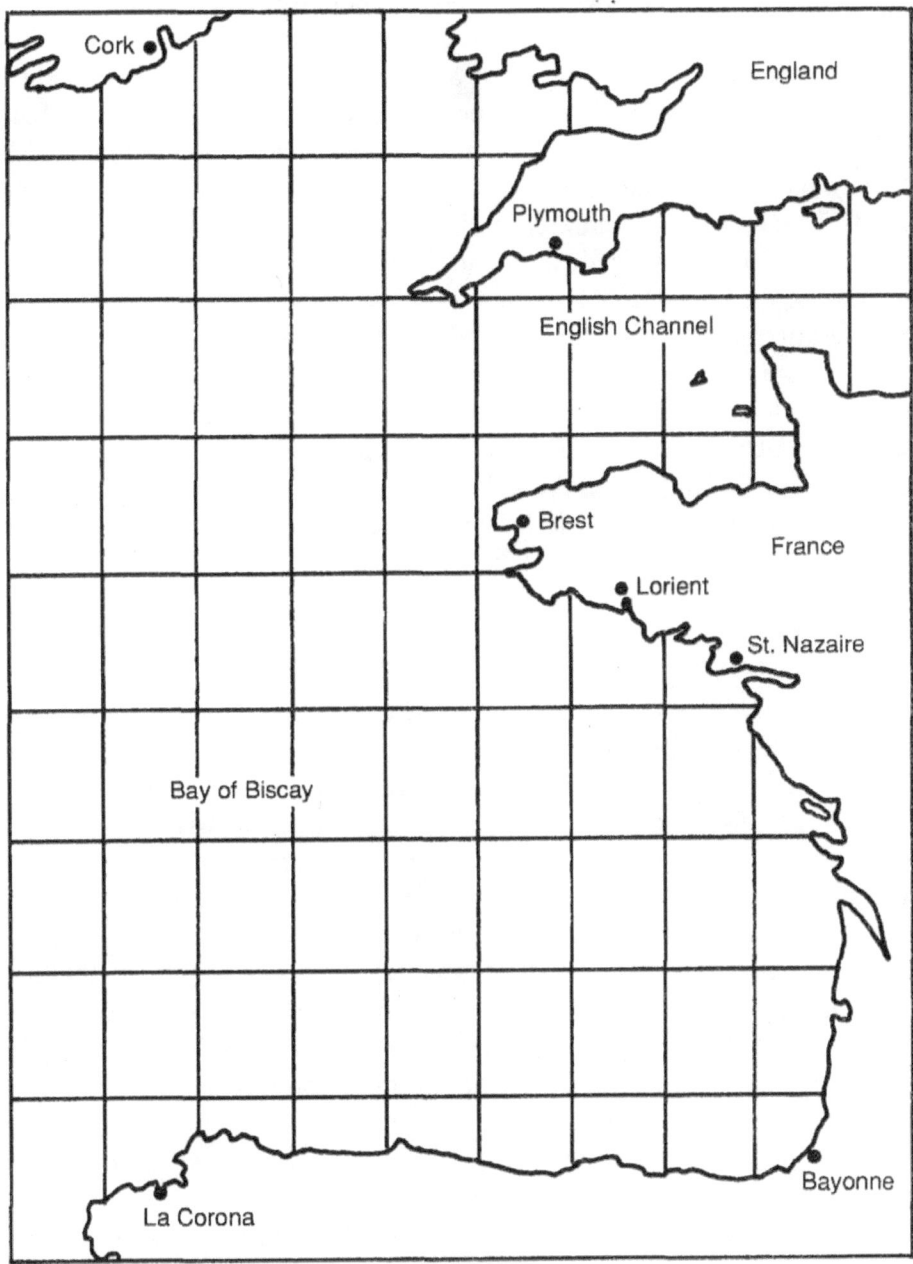

while holding as many variables constant as possible. At the start of this period the Allies began flying patrols around the clock—instead of just in daylight—setting the scene for an electronic confrontation of radars and radar countermeasures that effectively ended at the beginning of 1944, when the German introduction of the snorkel permitted U-boats to transit the Bay without surfacing to recharge their batteries.

Two principal forms of measure and countermeasure inter-played. On the hardware side, the British introduced a series of airborne radars (with the ASV (anti-surface vessel) Mark II and the ASV Mark III coming into use in the period considered here) with which to detect U-boats at night or through clouds. The Germans countered with a series of search receivers, notably the Metox and the Naxos, designed to detect British radars. Balloon-borne decoys proved ineffective and the Germans quickly abandoned them. On the operational side, the Germans adopted a wide variety of policies for transit of the Bay, ranging from lessening detectability by surfacing only at night to lessening vulnerability by remaining submerged at night and racing in packs across the Bay in broad daylight, the better to engage aircraft with FLAK (FLug Abwehr Kannonen) guns. British operations evolved as well, principally in lessening the incident radar energy needed to detect and approach submarines.

Approach

Quantitative examination requires measures of effectiveness—yardsticks with which to quantify success. Early operations researchers worked on the Bay of Biscay offensive and later used it as a textbook case in the creation of measures of effectiveness.[13] This book uses their measures as well as certain new ones. These measures reside in models: mathematical representations of reality. By using historical data in the models, we can measure the effectiveness of the two sides' operations in the Bay and perform various analyses of both the Bay operations and the U-boat campaign as a whole. The map of models and analyses sets forth the design of this study. (See figure 3.) The figure is to be read from left to right: the historical data in the left portion of the figure provide inputs to the models, which in turn permit the analyses. (Note that wartime analyses and models appear in *white,* while models and analyses created in this work are *shaded,* and that all data are necessarily of wartime vintage.)

After a qualitative summary of the course of Bay operations during 1942 and 1943, this study will use—as did wartime operations researchers—records of patrol flying hours, U-boat sightings, and U-boats present in the Bay to form the *operational sweep rate,** a measure of the Bay searchers' effectiveness. (See the upper region of figure 3.) Monthly readings of this rate provide time-series data we will use in several ways. The wartime version necessarily used intelligence estimates of the density of U-boats in the Bay; after reproducing that version we will repeat the process with authoritative U-boat data gleaned from Admiral Dönitz's *War Diary,* producing a somewhat different track record for the Bay searchers.

Wartime analysts gained considerable insight from the operational sweep rate time series without further mathematical ado, relating increases in search effectiveness to the introduction of new British equipment and declines in effectiveness to the introduction of new German equipment or tactics. Although wartime analysts recognized the possible presence of seasonal effects, they performed no seasonal adjustment. We will seasonally adjust the data in two different ways (one based on Fourier analysis and the other on linear regression), showing that the wartime researchers' conclusions were not artifacts of the confounding factor of seasonal change.

Working with the seasonally adjusted series, we will find its *autocorrelation,* a measure of the degree to which the level of performance in any given month coincides with the occurrence of the same level of performance in following months. Stemming from the autocorrelation analysis—and really making it worthwhile—is the analysis of how, on average, the operational sweep rate changed after an innovation. Though the rate does not in itself show any particularly strong pattern over the years 1942-43, it does show a distinct pattern of response to two "forcing functions": German and British innovations. Taking one forcing function at a time, we can distill from the sweep-rate data the response of the sweep rate to a single German or British innovation.

---

*   This term will be defined at some length later: it expresses the effectiveness of the searchers in terms of the number of square miles' worth of submarines they were finding. This concept proved very powerful and pervades the entire early literature of operations research, not only in obvious application to explicit search problems but in imaginative applications to such diverse problems as bombardment and decryption. Extensions of it are made, such as dividing the operational sweep rate of a platform by its speed, so as to obtain its operational sweep *width.*

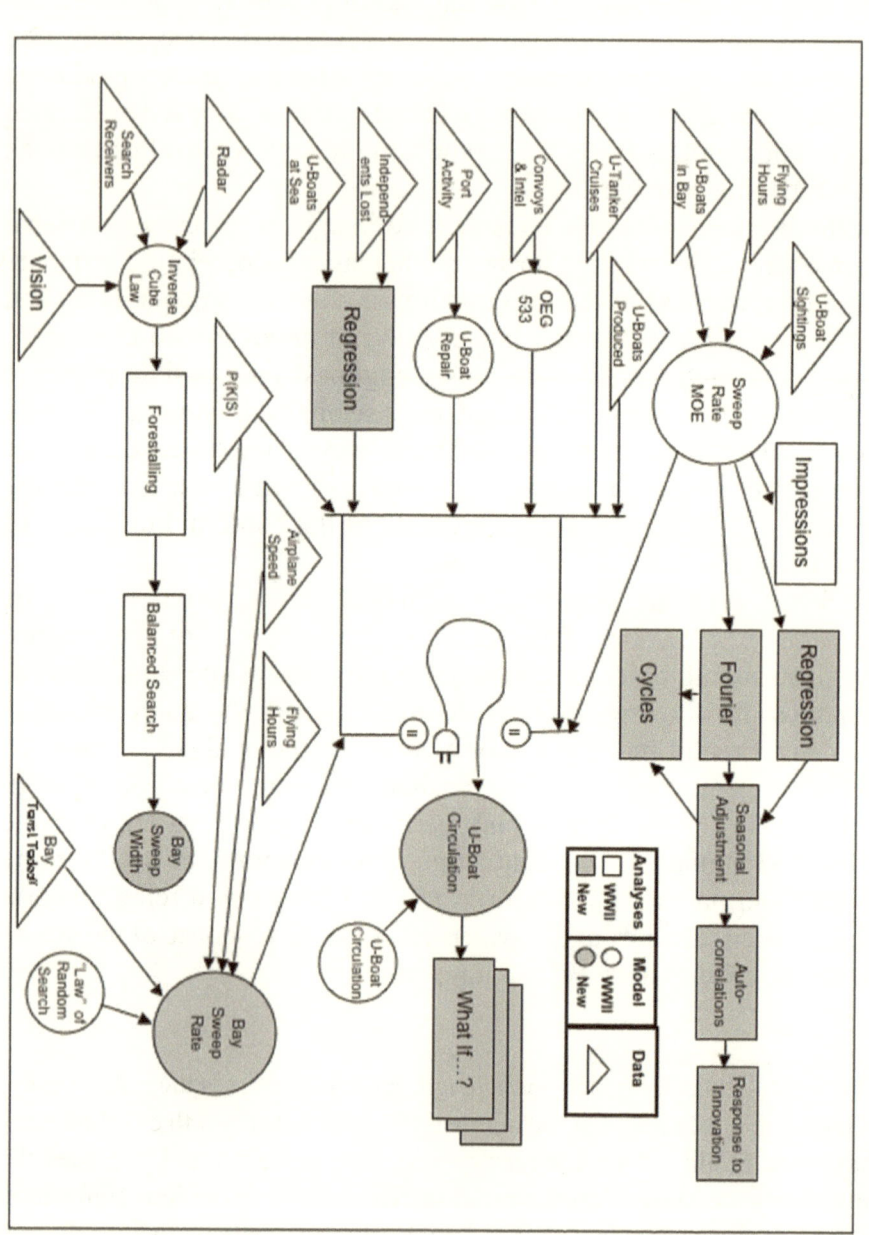

Figure 3
Models and Analyses Map

The investigations described so far resemble those of stock market "technical analysts" who study the patterns of price movements without considering the underlying characteristics of the stock-issuing companies. Such analysis can be useful, but it can go only so far. To understand *how* the different measures and countermeasures affected the sweep rate, we need a mathematical model of Bay search operations. To provide understanding, a model must be more than a least-squares fit of some arbitrary equation to the data. Such a "model" is a casting, not a sculpture: its mechanistic resemblance to the original enlightens neither the creator nor the beholder. Sculptor-like, we shall build up a model of Bay search through study of its anatomy, with our result not only identifiable as a likeness of the original but—through its very rough-hewnness—providing insight into the way the underlying skeleton of measures and countermeasures determines the visible outline, the measure of effectiveness.

We will begin (at the lower left of the Models and Analyses Map, figure 3) with some facts about radar and vision and the "inverse cube law": a simple model, created by wartime operations researchers, of visual and radar detection. This model provides an estimate of the probability that a submarine will be sighted by a passing airplane as a function of the distance of closest approach. Refinement of the model accounts for the probability that a lookout on the submarine, or the operator of a radar warning receiver, will sight the airplane before the airplane sights the submarine.

Airplanes could search, and U-boats could recharge their batteries, by day or by night. If the Germans restricted their surfacing to the safer nighttime hours, the British could restrict their search to the same period, opening up the possibility of totally safe daytime surfacing. We will reconstruct the analysis by which the British balanced their search effort, and the Germans their surfacing, between day and night. This balance changed over the course of each year as the amount of daylight varied.

Knowing the division of searching and surfacing between day and night, we need to know the absolute amounts of each for each of the 25 months under consideration. The amount of flying and the density of submarines are matters of record. However, we need a way to relate the total time a submarine spent in the Bay to the time it spent on the surface. Fortunately, this relationship—an inverse relationship because of the slow speed of submerged U-boats—appears in one of the postwar Allied sources and allows determination of the number of surfaced-U-boat-hours per month, given the total U-boat-days per month in the Bay and the total number of U-boats making passage. The resulting calculated degrees of submergence gain validity in that they show greatly increased submergence after Dönitz's issuance of his "maximum submergence" orders.

So far our approach has not addressed the operation of many aircraft in concert; we must capture mathematically the efficiency with which the Bay search operations were coordinated. Unlike the ranges of radar or visual detection, the speeds of aircraft and submarines, the length of the daylight portion of the day, or the optimum mix of daytime and nighttime operations, the degree to which search aircraft efficiently coordinated their efforts can be judged only by their results.

Although effectiveness has no upper bound, efficiency does: perfection. Therefore, we use equipment parameters and performance models to make two *a priori* estimates of the efficiency of coordination among aircraft engaged in Bay search operations. Each treats the result of air operations as the sowing of "lethal area" onto the Bay—although aircraft searched for U-boats by the square mile and then attacked them individually, we will summarize the process as one of exterminating submarines by the square mile. The first estimate will reflect the assumption that the aircraft coordinated their activities with perfect efficiency, never searching a portion of the Bay more likely to be sub-free than any other. The second estimate will follow the practice of wartime operations researchers, who usually made the pessimistic assumption that the search effort would be totally uncoordinated. The precise characterization of such a completely noncoordinated "random search"* as a Poisson distribution** of lethal area over the Bay will have a surprising application later on in the study. Historical values for the probability of kill given sighting—which rose markedly over the course of the war—enter into the calculation of lethal area.

Such a complete account of Bay search measures, countermeasures, and measures of effectiveness invites consideration of larger issues of effectiveness. Despite the emphasis on search operations, in which sightings are the goal, lethality and the probability of kill given sighting also figure in the analysis. These considerations will help answer questions such as, "For how many U-boat sinkings was the tardy introduction of the

---

*   As opposed to a perversely coordinated "bad search," which would deliberately
      concentrate on areas known to be sub-free.

**   A recipe's worth of chocolate chips are Poisson-distributed in the resulting
      batch of cookies.

Naxos receiver responsible?" by exercising the sweep-rate model with and without more timely introduction of the Naxos radar receiver in 1943.

Then another question arises: "How many Allied merchant vessels did tardy introduction of the Naxos receiver save?" To answer that we need to model the whole U-boat war in the North Atlantic. Our "U-boat circulation model," based on a system of ordinary differential equations developed by wartime operations researchers, will model U-boat operations as a flow of U-boats among several states, such as "at sea," "awaiting repair," and "sunk." Each month some proportion of boats in each state moves to other states. (Were the proportions constant over time, the model would be a Markov chain.) The U-boat circulation model occupies the center right portion of the Models and Analyses Map; notice that its Bay sweep rate input can be provided either by the historical sweep-rate data on the upper left or by the output of the Bay sweep-rate model on the lower left, In either case, the model relies on the data and analyses in the left center portion of the figure.

Wartime operations researchers developed a simple U-boat circulation model, highlighting the repair step of U-boat circulation in order to understand the tradeoff between using aircraft for antisubmarine patrol and using them for bombing repair facilities. We will use their U-boat repair submodel, whose formal resemblance to the "law of random search" furnishes some insight into the nature of the repair process. By focusing on that process, wartime researchers lumped U-boat construction and sinkings into a single constant, the change in U-boat strength per month. However, because we seek to understand the effect of Bay operations upon the whole U-boat campaign, and in particular upon the number of merchant vessels sunk, we must treat these elements of the process separately.

Sinkings of U-boats on the high seas resulted almost entirely from encounters with Allied ships and aircraft escorting convoys of merchant ships. Considerable analysis of such encounters was done during the war, providing almost ready-made models of the U-boat search for convoys and for the expected sinkings (U-boat and merchantman) that resulted. These analyses also provide estimates of the convoy densities and levels of German benefit from decrypting of Allied radio messages over the period of interest.

A tanker U-boat could refuel other U-boats at sea, giving about 10 boats an extra month's endurance each. The wartime operations researchers knew the average duration of a U-boat cruise but they did not include the tankers as a determinant of U-boat endurance at sea. Because we now know the schedule of at-sea refuelings, we can include it in our model, along with historical values for the number of new U-boats introduced per month. Note that the usual cost of adding explanatory variables—the concomitant addition of degrees of freedom in the model—does not occur here because these explanatory variables do not carry with them undetermined coefficients to be "fit" by such a method as least-squares: a newly built boat sailing from Germany results in exactly one new boat entering the North Atlantic; operating a U-boat tanker results, on average, in 10 boats staying there.

Indeed, we use multiple regression only to estimate the number of independent (that is, nonconvoyed) merchant vessels sunk as a function of U-boats at sea and of months since the entry of the United States into the war.* Strictly speaking, a U-boat circulation model would not need to address the sinking of independent merchant vessels because such vessels could not fight back and thus did not affect U-boat circulation.** However, we want to measure the effectiveness of Bay search in terms of the total number of merchant sinkings prevented, so we need to model the sinking of independent merchant shipping. The regression analysis, with a satisfactorily high R^2, provides a function with which to do so, but despite the definition of R^2 as "percentage of variance explained," the coefficients determined by the regression explain nothing and do not even have identifiable physical interpretations, thus standing in contrast to the speeds, sighting distances, probabilities, and other parameters used elsewhere in the analysis.

Once complete, the U-boat circulation model enables assessment of the effectiveness of different Bay measures and countermeasures, using merchant ships saved (or sunk) as a measure of effectiveness. We can use either the historical values of operational sweep rate in the Bay or those

---

*   Single-variable linear regressions, moving averages, and the like will be used to smooth such series as the U-boat/merchant vessel exchange rate and the probability of kill given that a sighting has taken place. These instances do not detract from the point at hand, which is that regression results aren't compelling models. Precisely because we do not undertake to model tactical prosecution of sightings (of either U-boats or convoys), we use linearized or smoothed historical values.

**   In principle they could draw out more U-boats, but in fact Dönitz kept as many U-boats at sea as possible, except perhaps for a brief period in 1943.

values derived from the sweep rate model. Using the latter, we can address various "What if . . . ?" questions regarding search operations in the Bay of Biscay and the U-boat war in the North Atlantic, such as "What if there had been no air patrol in the Bay of Biscay at all?"

Results

The Dönitz data portray the search effort as less productive of sightings than the wartime researchers had thought. In 1943 especially, researchers were unaware that the U-boats had begun to spend less time on the surface while making Bay passage; submerged boats traveled more slowly than surfaced ones, so the analysts were simultaneously undercounting boats in the Bay and overestimating the number on the surface at any given time. Two discrepancies appear between the performance according to the Dönitz data and according to Allied wartime intelligence. The Dönitz data show phenomenal success in January 1942 and appear to show that nighttime radar search of the Bay (which required the use of a special searchlight, developed contrary to orders, to make the final approach to the target) in fact began a month before its official June 1942 start. The success of January 1942 does not appear in the Allied version because Allied U-boat density figures, arrived at separately from sightings, portray U-boats as especially plentiful that month when in fact they were not. January 1942 *was* a good month for decryption of the German Enigma cipher; one may hypothesize that unwitting searchers received orders to search in areas particularly likely to contain U-boats, and that the operations researchers received inflated figures for overall U-boat density in the Bay so that they would not realize, or reveal in their results, that a unique source of U-boat position intelligence existed!

The Fourier analysis shows—as expected—a strong seasonal component and, as hoped, that the fluctuations remaining after seasonal variation has been removed correspond to the introduction of new equipment by the two sides. In fact, two nonseasonal components appear, which we may tentatively identify as a "British cycle" and a "German cycle."

The autocorrelation analysis shows that blocks of five consecutive months are marginally correlated; beyond a five-month span the correlation breaks down completely, presumably because of the large chance that one side or the other would have introduced a device or countermeasure during that time. The autocorrelation analysis also enables us to find the average response of the sweep rate to the introduction of an innovation by one side or the other. Interestingly, the responses—though immediate—include a backlash to the disadvantage of the innovating side, starting about six or eight weeks after the innovation. The sweep rate then displays damped oscillation over many months, showing that neither side could gain the upper hand permanently.

The "random search" estimate coincides closely with the operational sweep rate (square miles' worth of submarines sighted) as computed from the *War Diary* data, except that the estimate does not reproduce the great Enigma success of January 1942 (nor should it, as Enigma does not figure in the model at all), and it does show the huge surge in sweep success associated with night radar search as occurring in June 1942 instead of May. In these two respects the sweep-rate model's values for the operational sweep rate more closely resemble the wartime researchers' values than the *War Diary*-derived values.

From the "What if . . ." analyses emerges a detailed understanding of the relationships inherent in the U-boat campaign. As Dönitz realized even before the war began,[14] the U-boats, despite their submersibility, moved and fought best on the surface and submerged only when threatened: they could be thought of as submersible torpedo boats. Bay search turns out to have been effective largely because it encouraged U-boats to make lengthy submerged passages, cutting into the time they could spend at sea sinking merchant vessels. To extend that time, the Germans started a program of at-sea refueling using tanker submarines; this program had the important side effect of deferring maintenance, alleviating a severe backlog that clogged the occupied French ports from which U-boats operated. The Allies, in an unusually overt use of information gained from decrypting German radio traffic, sank these tankers. Thus, tanker U-boats can be shown to be the most effective aid to Bay transit, and Enigma decryptions can be shown to be the most potent tool of Bay search.

The German maintenance backlog of U-boats in France grew throughout the war. Had the Germans increased their maintenance capability, they would have sunk many more merchant ships than they did. As it was, if a boat returned to France, it incurred a measurable and increasing risk of spending the rest of the war waiting for repairs and never putting to sea again. Boats staying at sea through at-sea replenishment could avoid the round-trip passage through the Bay, with its attendant risk and opportunity cost in time not spent searching for convoys. A subtle benefit of the U-boat tankers was their ability to postpone the entry of boats into the repair quagmire.

The Limits of Quantitative Analysis

Many balk at the use of quantitative methods in the study of military matters, holding that judgment and experience—perhaps their own—have far more to offer than any calculation possibly could. Judgment and experience *are* valuable guides in any human endeavor. However, war—and especially the kind of search operation investigated here—entails a great deal of uncertainty. Those who deal with uncertainty in war by playing hunches seem likely to share the fate of those who use that approach in poker or backgammon. In fact, the wartime operations researchers pioneered the use of quantitative methods in military matters, and their success convinced many uniformed skeptics that in fact "the long-haired boys can help."

Two common objections to quantitative analysis of national security matters boil down to claims that such analysis takes an oversimplified view of the participants. The first is the assumption that some group can be considered a "unitary actor." This book contains many references to "the Germans" as if they constituted a monolithic unitary actor. Political scientists and economists have found that a putative unitary actor (such as "the household") often turns out to be a barrel of monkeys best understood individually.[15] In the context of U-boats, however, "the Germans" really means Grand Admiral Karl Dönitz. The next lower authority was the individual skipper, whose daily and sometimes hourly reporting requirements and new orders left little doubt as to who truly directed his submarine.[16] Dönitz thus makes a tempting target for "What-if" questions (traditionally ruled out by historians because they proceed from a "counterfactual," i.e., false, premise): we need posit only counterfactual behavior for a single individual. The Allied flying effort was far less unified, but a quantitative analysis can take that into account, as outlined above, through the concept of "random search."

The second alleged oversimplification is the treatment of the participants as "rational actors" bent on reciprocal frustration, when in fact they suffer from the human shortcomings of inde-cisiveness and fallibility and pursue only tangentially conflicting objectives.[17] Whereas those who identify actors as nonunitary advise breaking them down into true units, those who decry the assumption of rationality do not offer a particular assumption of irrationality with which to replace it. In the case at hand, however, we may cite historical evidence as to Dönitz's calculating rationality and we will devote considerable mathematical effort to dealing with human fallibility in the crews of aircraft and submarines. Crucially, Dönitz and the Allies operated according to the same objective function: merchant vessel sinkings. As Dönitz's *War Diary* states:

> The enemy powers' shipping is one large whole. It is therefore immaterial where a ship is sunk . . . . Tonnage must be taken where it can be destroyed most reasonably as far as making full use of the boats is concerned.[18]

American analysts agreed:

> To the extent that . . . the enemy [is] prevented or hindered from transporting necessary cargoes, the submarine offensive makes a major contribution to the progress of the war. In the submarine's antishipping offensive three things must be accomplished: achieving contacts on ships, approach to within torpedo range, and final attack. (. . .) The primary aim of ASW [anti-submarine warfare] is to reduce the effectiveness with which the submarines carry out these steps, and the success of an antisubmarine effort is to be assessed in these terms. Antisubmarine warfare is not an end in itself, but merely a means of ensuring that the ability to use the ocean is maintained at the best level possible.[19]

Even given a valid premise of rational and determined adversaries seeking diametrically opposed goals, any quantitative analysis—including that presented here—inevitably contains many approximations, simplifying assumptions, shortcuts, and discrepancies between calculated and observed quantities. One who had low esteem for quantitative analysis of military affairs could accept the present work as correctly done but cite these as signs of the limitations of a quantitative approach. Yet other approaches, such as the traditional historian's narrative account and lessons learned, also begin and end with something less than the whole truth. The difference between the traditional approach and a quantitative one is that in the former most departures from the ideal can be discerned only by considering external evidence; a quantitative approach—*if applicable at all*—can sound the depth of its own shortcomings.

A Note on Sources

Three American sources deserve special mention not only for their outstanding quality but because they have provided many of the building blocks used in assembling the present work. All are immediate postwar compilations of wartime Operations Evaluation Group studies, and will amply repay any amount of study.

Operations Evaluation Report (OER) number 51, *Antisubmarine Warfare in World War II* by Charles M. Sternhell and Alan M. Thorndike, presents a narrative history of the war against the U-boats, as well as the analytical methods early operations researchers used to improve the Allies' antisubmarine effectiveness. Chapters 13 and 14, "Offensive Search" and "Employment of Search Radar in Relation to Enemy Countermeasures," provide technical and operational data and an introduction to the operational sweep rate.

OER number 52, *Methods of Operations Research* by Philip M. Morse and George E. Kimball, uses offensive search of the Bay of Biscay as measured by operational sweep rate as an example of a successful choice of a measure of effectiveness. Morse and Kimball also present the simple U-boat circulation model created during the war. Though simpler than the one used here, their model and discussion provide some useful data and flag U-boat circulation as a topic for investigation.

OER number 56, *Search and Screening* by Bernard Osgood Koopman, is the most strongly mathematical of the three books. Deservedly brought back into print in 1980 by Pergamon Press, it contains the rigorous mathematical underpinnings of the work in the other two books. This study's calculation of sweep rates of detectors in the presence of counterdetectors draws several concepts and one characteristically elegant solution from Koopman.

These American sources are complemented by C.H. Waddington's *O.R. in World War 2*, written immediately after the war from the author's personal experience in the British antisubmarine operations research effort, but not cleared for publication until 1973. From the other side, Grand Admiral Karl Dönitz's *War Diary* (also known as the "BdU Log," reflecting Dönitz's title, *Befehlshaber der Unterseeboote)* provides a day-by-day record of U-boat locations as well as a wealth of commentary and several complete reports included for the record.

2

Qualitative Synopsis of Action in the Bay

From the moment that it had started in May 1940, the advance of the German Army in the campaign against France had been watched with close attention by U-boat Command. If the army succeeded in defeating France we should be given the advantage of having bases on the Channel and Biscay coasts for our naval operations against Britain. This would indeed be a sudden realization of our hopes for an improvement in our strategically unfavorable geographic position vis-a-vis Britain; it would mean that we should now have an exit from our "backyard" in the south-eastern corner of the North Sea and would be on the very shores of the Atlantic, the ocean in which the war at sea against Britain must be finally decided. The danger that enemy measures on a grand scale might prevent the U-boats from putting to sea would no longer exist, for such measures, if possible at all, could only be carried out in the shallow waters of the North Sea. Moreover, with bases on the Atlantic the distance which the U-boats would have to cover in order to reach the main British trade routes would be materially shortened, and even the small 250 ton Type II boats would then be able to operate in the Atlantic. In addition new repair yards would become available to us, the dockyards at home would be relieved of the burden of overhauling existing

boats and they could concentrate on the building of new vessels.
All in all, possession of the Biscay coast was of the greatest
possible significance in the U-boat campaign. Once we had
that coast, there was only one task on which the German Navy
had to concentrate its efforts, the task of taking, as swiftly as it
could and by every means it possessed, the greatest and most
comprehensive advantage of this outstanding improvement in
our strategic position at sea.
            —Grand Admiral Karl Dönitz, 1958[20]

Starting in September 1941, British aircraft patrolled the Bay of Biscay, searching for U-boats in transit to and from their ports in occupied France.[21] The effort was termed "offensive" search as opposed to the "defensive" search for U-boats in the immediate vicinity of convoys. The idea was to create a so-called "unclimbable fence,"[22] a patrolled region so wide that a U-boat could not complete the transit underwater: when the boat rose to the surface to run on diesel—rather than electric—power and recharge its batteries, it would be vulnerable to attack from the air. (The German Navy had yet to adopt the snorkel, invented before the war in Holland.) By December the U-boat force had learned to surface only at night, and British forces had countered with the introduction of a radar, the Anti-Surface Vessel (ASV) Mark I.

The Leigh Light Enables Nighttime Radar Search for U-Boats

With the equipment available at the beginning of the Second World War—or even at the end—an aircraft could not home in on a surfaced submarine by radar alone.[23] Not only did the submarine cease to stand out from the sea surface as the aircraft came overhead, but the airplane's radar could not switch from its role as a transmitter to its role as a receiver quickly enough to catch the echo from a submarine nearer than about a mile. Thus the bombardier needed a second sighting device for the final approach in nighttime bombing runs against surfaced submarines.

Squadron Leader Humphrey Leigh heard of this problem in the autumn of 1940 and, operating outside official channels, developed a large carbon-arc searchlight for installation on Wellington bombers on submarine patrols. The scientific establishment was content with its system based on flares and a smaller light. Leigh's commanding officer, however, allowed him to develop a prototype of his light. Although it proved successful in trials against a British submarine, the military establishment faulted it in favor of *their* system, based on a third, even larger, light. But this light, unlike Leigh's, was aimed by moving the whole airplane. It also tended to dazzle its user because of its great power and its location in the very nose of the airplane; Price points out that the Leigh Light, mounted on the belly of the plane, prevented dazzle (as do low-mounted fog lights on a truck), because the pilot looked over the beam, not through it. After two months, Air Chief Marshal Sir Philip Joubert admitted that his opposition to the Leigh Light had been a mistake and reversed it: in August 1941 Leigh received authorization to turn his prototype into a producible weapon.[24]

Records show that Leigh Light-equipped Wellington bombers first flew night patrols in the Bay of Biscay in June 1942.[25] These Wellingtons carried the ASV Mark II radar, a set capable of detecting submarines at ranges of up to 10 miles, much farther than the Mark I. Like the Mark I, it sent and received electromagnetic waves about a meter in length.

Ground Rules of the Electronic Contest

What one receiver can detect, another can. Radar works by bouncing radio waves off the target and detecting the returning echo. A radar warning receiver—carried by the target—can detect a radar signal as it bounces, providing warning of radar surveillance. Motorists' "fuzz busters," used to detect police radar speed traps, are radar warning receivers.*

The problems facing the designer of a search receiver relate to those facing the designer of the receiver part of the radar itself in interesting ways. Fundamentally, each strives to extract the radar signal from the background "noise": signals of other radio and communications gear, the inadvertent electromagnetic radiation of electrical equipment such as motors, and naturally occurring electromagnetic waves. The radar designer has the advantage of knowing exactly what signal to seek: it will be the echo of the one sent out earlier. Hence he can tune his receiver to the precise

---

*   Radar warning receivers were called "search receivers" by the wartime operations researchers, a usage adopted for the balance of this study.

frequency of the expected signal, tuning out great amounts of noise. He can also be sure the echo will come from the direction in which the radar's dish points: the directionality of the dish assures reception of the entire signal and only a fraction of the noise.* The designer of the search receiver, on the other hand, cannot rely on exact knowledge of the radar frequency, or of the direction from which the radar signal will arrive. However, he enjoys an advantage over the designer of the radar itself in that the radar energy striking the target far exceeds the energy finding its way back to the aircraft's radar. The signal must make a round trip to get back to the radar, and it spreads out on each leg of the trip. The spreading on each leg weakens the signal in proportion to the square of the distance traveled. Hence the search receiver need only detect signals whose intensity obeys an inverse-square law of attenuation with range, whereas the radar itself must detect echoes that have diminished according to the fourth power of the range. These factors usually balance out somewhat in favor of the designer of the warning receiver.** However, some means of separating signal from noise must still be used. The Germans had two methods at their disposal: the heterodyne circuit, which preferentially amplifies signal over noise but emits radio waves of its own, and the use of some kind of directional antenna. Each of these approaches has its drawbacks but each, as we shall see, was used.

In U-boat hunting, however, an operational consideration—for both antagonists—tended to compensate for the inherent physical advantages of the search receiver. This consideration was the tradeoff between false alarms and misses.

Almost any detection device, be it an ASV radar, a radar warning receiver, or a household smoke alarm, converts a measurement into a binary (yes-no) decision. Some detectors, such as smoke alarms, accomplish this conversion on their own, whereas others, such as radiation film badges, merely act as transducers, converting whatever they detect into a form a human operator can perceive. In the former case, the device is set so that a measurement beyond some threshold value will trigger the alarm. In the latter, the operator chooses his or her own threshold.

---

*   This is the effect one gets by cupping a hand to an ear. The statement disregards antenna sidelobes.

** Hence some states have resorted to declaring "fuzz busters" illegal, applying a legal remedy in the absence of any technical one.

Whether the threshold is built in at the factory or chosen by the operator, the level at which it is set embodies a choice between missing some detections and getting some false alarms. A detector with a lower threshold will miss few detections and will give many false alarms; a detector with a higher threshold will give few false alarms but will also miss more genuine detections. No change of the threshold setting can simultaneously lessen both the false-alarm rate and the miss rate. The relative costs of the two types of error determine the correct setting, which is the one that minimizes, over the long run, the total cost inflicted by both types of error. Using the example of a household smoke detector, a missed fire would be a disaster while a false alarm is only annoying, so the householder accepts a threshold at which he or she will hardly ever miss a fire but will hear occasional false alarms.

In the case of an Allied aircraft using radar to detect a surfaced U-boat, which in turn is using a search receiver to detect the aircraft radar, the two operators face different relative costs of detections and false alarms. The radar operator can ask his pilot to investigate a possible sighting with little fear of adverse reaction if no submarine is found—the airplane has to fly patrol all night in any case.* The search-receiver operator, on the other hand, will think twice before advising his skipper to stop recharging batteries and crash-dive the boat: the risk of attack from the air must be balanced against the necessity to recharge batteries and the need to complete the transit quickly.

Metox, an Expedient Countermeasure

The Germans, having removed a Mark II set from a bomber that crashed in Tunisia in the spring of 1942,[26] deduced the existence of British meter-wave radar almost as soon as the British used it against them in an antisubmarine mode.[27] The Germans quickly adapted and deployed an existing French receiver, dubbed Metox after one of the firms that produced it, to detect the British radars. A submarine equipped with Metox could detect a Mark II radar much farther away than the radar could detect the

---

*   Because of the low cost to the aircraft of prosecuting false alarms, one would not expect decoy targets resembling submarines to degrade the search performance of the aircraft to any significant degree. Indeed, the Germans used such decoys but with little success.

submarine. However, the heterodyne-based Metox had the then-overlooked flaw of emitting strong radio waves itself, in principle permitting passive detection far beyond radar detection range.[28] Metox receivers went into use in September 1942,29 and though some boats still did not have Metox by the end of the year, transit in groups including boats that did have Metox allowed all to benefit from the invention.

The Germans, including Hitler himself, were wildly enthusiastic about the Metox receiver's effectiveness.[30] Outside the Bay of Biscay, however, this receiver did little good, leading Allied operations researchers to attribute—at least in hindsight—decreased Allied search efficiency in the Bay to some other factor.[31] Although wartime operations researchers thought that seasonal and psychological factors could be at work, no attempt to adjust patrol data seasonally appears in the literature.* The particular psychological effect considered was overreaction to the Leigh Light: the submarines began the dangerous practice of surfacing in the daytime, until confidence in Metox—though largely misplaced—caused a return to the safer practice of surfacing only at night.

A New Airborne Radar Is Deployed

The Mark II's decreasing usefulness and other considerations impelled the British to develop a radar that would operate on a new, shorter, wavelength.[32] The resulting S-band Mark III radar had a wavelength of about 10 centimeters and first went into use in the Bay of Biscay in early 1943, not quite three years after the invention of the enabling technology, the strapped magnetron.[33] The Metox receiver could not detect the Mark III at all because the 10-centimeter wavelength placed it outside the spectrum to which the Metox listened. Radar search in the Bay proved increasingly successful through midyear.[34]

The Luftwaffe recovered a Mark III radar from a crashed airplane, albeit one not used for antisubmarine patrol, almost as soon as the radar was introduced.[35] Despite having the ASV Mark III, called the "Rotterdam Gerät" by the Luftwaffe after the place of its discovery, and despite the clearly increased success of Allied patrol aviation, six months elapsed before the Germans connected the captured centimeter-band radar with their rising submarine losses.

---

* Such an adjustment will be made later in this study.

Two diametrically opposed nonelectronic countermeasures helped protect the U-boats during this period. In April 1943 Dönitz mandated the seemingly paradoxical policy of surfaced transit of the Bay by day and nighttime submergence.[36] This policy amounted to a concession that detection by patrol aircraft would be automatic, day or night. By day, however, the U-boats could in turn use their FLAK antiaircraft guns to fight back. This idea—about which Dönitz felt so enthusiastic that he issued a radio message ordering its adoption by boats already at sea—did not work because the FLAK guns were not effective enough against the heavy patrol aircraft. In June Dönitz added a refinement: the submarines started making the transit in groups of three to five, relying on the fact that patrol aircraft carried only enough ordnance to destroy a single U-boat. Again, Dönitz radioed the new policy to his U-boats for immediate implementation.[37]

The British vitiated the group-transit tactic by flying aircraft in loose formation, so that any one plane coming across a group of submarines could summon the others.[38] Because the pack method provided insufficient safety for transiting submarines, Dönitz soon ordered a policy of maximum submergence at the expense of greatly increased transit time. Dönitz had tried maximum submergence with night surfacing for a while in 1942 and rejected it; the 1943 policy incorporated the new concepts of pack operation and surfacing by day. The new policy worked very well, but Dönitz decided to configure some U-boats specifically as FLAK boats: they would do nothing but parade about the Bay in the hope of attracting patrol aircraft, which they could then shoot down. Results with the first such boat were disappointing and Dönitz eventually abandoned the idea.

The U-boat force also tried Aphrodite, a radar decoy consisting of a balloon and strips of foil. This countermeasure did not work well, probably because the aircraft paid no great penalty for investigating a false sighting.

In another effort to enhance counterdetection of aircraft by submarines, the Hohentwiel aircraft-warning radar entered service in the second half of 1943.[39] The ASV Mark III outranged the Hohentwiel, and in any case the U-boat commanders hesitated to use it for fear that Allied planes could detect its emanations. Although most airplanes could not, the few "ferret" planes dedicated to the detection of U-boat radars found that use of the Hohentwiel was rare.

Naxos-U Throws the Germans Off

The Naxos-U device, introduced in May 1943, could, in principle, receive the 10-centimeter wavelength of the Mark III radar.[40] In practice, however, the device was so insensitive and awkward to use that it could not detect the signals of the ASV Mark III radar at all, which deceived the Germans into believing that no such radar was in use.

During this period the Germans entertained several incorrect explanations of the mounting danger to their submarines, which was in fact due to the Mark III radar.[41] They added a new fourth wheel to the U-boat version of the Enigma machine to enhance the security of U-boat communications. They wrongly suspected the British of using an infrared submarine-detection device. When the British learned of this suspicion they planted false confirmations of it to heighten the confusion. Paints intended to reduce the infrared signature of the submarines actually increased their radar signature,[42] and finally the Germans blamed emissions of the Metox receiver itself for the ease with which the British could find U-boats. Not only did experiments reveal that a Metox could be detected at a range of 30 miles, but a British prisoner daringly "revealed" that Allied aircraft homed in on the Metox sets from fabulously long ranges.[43] In early August Dönitz radioed all U-boats to stop using their Metox sets.[44]

The "Wanz" (short for *Wellenanzeiger*) receiver, also called "Hagenuk" after its manufacturer, appeared at this time but saw little use because the U-boat skippers feared even its greatly reduced level of emissions. In any case, the spectrum of signals it could receive did not include wavelengths as short as 10 centimeters. The Borkum receiver followed. It did not use the offending heterodyne method of signal enhancement and therefore did not radiate at all, but it too failed to cover the 10-centimeter wavelength.

An Improved Naxos Finally Detects the Mark III

In November 1943 a Wellington crashed in France with a Leigh Light, a depth charge, and a Mark III radar on board, finally clinching the Germans' September realization that the "Rotterdam Gerät" was in use as an antisubmarine search radar. Even without this knowledge, the Germans had improved the Naxos, making it marginally effective against the Mark III radar by October 1943. In early January 1944, the Germans fitted out a U-boat with a variety of electronic and infrared equipment in a systematic attempt to deduce what detection methods the Allies were using. The boat was sunk by British surface vessels in February, with Dr. Karl Greven—a German radar scientist and U-boat officer—taken prisoner.* Truly effective S-band German search receivers would not appear until April 1944.[45]

Allied aircrews had started to take tactical precautions against a possible S-band search receiver long before the deployment of either the ineffective Naxos-U or the improved Naxos. Sternhell and Thorndike list four such precautions:[46]

1) Maintaining a normal radar scan, as opposed to holding the radar on the target during the approach, to foster the impression aboard the U-boat that a radar-equipped aircraft, though present, had not detected the boat.
2) Using an electronic add-on device (such as "Vixen") to turn down the radar's power during the approach to keep it below the threshold considered significant by the search-receiver operator.
3) Intentionally aiming the radar beam slightly away from the target during the approach, again to avoid strong illumination of the target.
4) "Turning the spinner [radar dish] aft . . . and approaching by dead reckoning."

Sternhell and Thorndike characterize the last precaution as "not very promising" because of the difficulty of making an unaided approach. They may have misunderstood the purpose of turning the spinner aft instead of shutting it off: many radars project some energy out a "back lobe" opposite to the main beam, and the aft-pointed radar would thus function as a forward-pointed radar of very low power. This speculation is suggested by the second and third precautions and deemed likely by a former wartime radar scientist.[47]

---

*   Both Morse & Kimball (p. 96) and Sternhell & Thorndike (p. 157) say that other experimental submarines had short careers. Knowing of the Allied success in breaking Enigma, one cannot help but wonder if signals intelligence led to the brevity of these careers or to any unusual efforts to take prisoners from Dr. Greven's U-boat.

The Race to X-Band and the Beginning of Stealth

When a bomber carrying an X-band radar, which operated on even shorter wavelengths (3 centimeters) than did the Mark III, crashed in Berlin in January 1944, the Germans immediately started work on a search receiver—to be called Tunis—designed against it, sure that sooner or later the radar would be used for antisubmarine work. To achieve adequate separation of signal from noise and yet avoid the emanation-producing heterodyne circuit, the designers of the Tunis search receiver used a directional antenna. Of course, the antenna had to be able to rotate because an airplane could approach from any direction. Allied aircrews knew that the Tunis swept a full 360 degrees about twice per minute, and countered by activating their radars only intermittently.[48]

The snorkel allowed the submarine to take in air while at periscope depth, permitting it to run on diesel power with only the head of the snorkel exposed. Snorkels first appeared on operational German submarines in early 1944.[49] A snorkel's maximum radar-detection range was about a third the range of a submarine, and thus close to the clutter-limited minimum range of a radar against any target at sea. In addition, antireflective snorkel coatings developed by the Germans further reduced the radar-detection range.[50] The Bachem net, a less successful device intended to reduce the detectability of surfaced submarines, underwent experimental use at about this time. The net consisted of a network of wires held one-quarter of a radar wavelength away from the U-boat hull, so that outgoing reflections from the hull and the shell of wires would cancel each other out through destructive interference.[51] Tables 1 and 2 summarize the types of radars and search receivers used in the Bay of Biscay and the chronology of their use.

Broad-Based German Solutions to the Bay of Biscay Problem

Tanker submarines,* originally invented for use in such distant waters as the South Atlantic and the Indian Ocean, began to resupply submarines engaged in North Atlantic anticonvoy operations in the second half of 1942.[53] At-sea replenishment of German submarines effectively amounted to a countermeasure to the Biscay offensive because each refueling eliminated the need for two trips through the Bay.

---

* U-boat types XIV and X B.

Table 1.—Summary Table of Radars and Search Receivers[52]

Band* (cm)	Radar	Nominal Range vs. U-Boat (miles)	Receiver
L (150)	ASV Mk I	4	—
L (150)	ASV Mk II	10	Metox
S (10)	ASV Mk III "H₂S"	15 or less	Naxos
X (3)	ASV Mk X "H₂X"	32	Tunis

(H_2S, H_2X)

Table 2.—Summary of Events in Radar War

February 1942	German "Enigma" Cipher Upgraded
June	Night L-Band Radar Flying Authorized
July	First Maximum Submergence Order
July	Metox GSR Deployment Begins
December	Metox Fully Deployed
February 1943	S-Band Radar Deployment Begins
April	S-Band Deployment Complete
April	U-Boats Fight Back with FLAK
May	Naxos-U GSR Tried Unsuccessfully
July	U-Boats Maximize Submergence Again
August	Wanz and Borkum GSRs Ineffective
September	S-Band Threat Recognized
October	Effective Naxos GSR Deployed
December	Allies Learn of Naxos
January 1944	Work on Tunis X-Band GSR Begins
February	Snorkel Deployed

Overall, one could see any step taken to improve U-boat effectiveness as a countermeasure against any antisubmarine measure then in effect. In particular, three major non-electronic hardware improvements—the Walter closed-cycle propulsion system, the snorkel, and the Zaunkönig (or "GNAT"—German Naval Acoustic [homing] Torpedo) qualify as countermeasures to the Bay of Biscay offensive on empirical grounds.

---

*   The letters used to denote portions of the spectrum are those used during the war, not those in use today.

Only the periodic need to use diesel power to recharge the batteries for underwater propulsion obliged U-boats to surface while in the Bay. Trying to alter the rules of the game, the Germans ordered 180 sea-going, Walter closed-cycle U-boats in June 1943.[54] Earlier work and a small prototype had validated Helmuth Walter's idea of using hydrogen peroxide instead of atmospheric air as a source of oxygen. His prototype submarine could run indefinitely on diesel oil underwater, but the attempt to scale up to the sea-going version proved premature. In the meantime, the Germans worked on the snorkel, which would allow underwater operation on diesel limited only by the need to remain at periscope depth.

In an even more basic countermeasure Dönitz cut back on U-boat operations altogether in August 1943, recalling some boats that had just set out, delaying the departure of others, and requiring those that did travel through the Bay to do so in Spanish territorial waters, surfacing only at night. This period of retrenchment ended a month later when improved FLAK guns, the Wanz ("Hagenuk") search receiver, and an acoustic homing torpedo were ready.[55]

Tanker Submarines and the Role of Enigma

As became well known in the 1970's, the Allied side benefited greatly from decryption of German radio messages enciphered on the military version of the Enigma multiwheel cipher machine. Although U-boat densities justified only "defensive" search in the vicinities of convoys and "offensive" search in the Bay of Biscay, the use of Enigma information enabled another type of antisubmarine operation: the interruption of at-sea resupply. Such resupply took place on the high seas, where the vastness of the open ocean and the comparative rarity of the event would make detection by air patrol extremely unlikely. Therefore, exploitation of any Enigma-based knowledge of at-sea refueling plans might have served as the textbook example of an operation forbidden on the grounds that it could reveal the Enigma source to the Germans.[56]

However, in June 1943 the USS Bogue used Enigma and other information to sink a tanker U-boat just before a mid-ocean rendezvous. The German reaction to the ensuing calamity—revealed by further Enigma decryptions—raised Allied estimates of the efficacy of such attacks so much that Enigma information was used freely by the Americans in an outright offensive against the refueling locations.[57] Although this offensive sank several submarines, tanker and otherwise, sources differ as to the utility of the Enigma information employed.[58] Indeed, the official British history of the operational use of intelligence in the Second World War cites the Enigma contribution as hard to assess and of mainly indirect value.[59] According to this interpretation, decryption took so long that a broken message could not be used to vector an aircraft to a U-boat, but the copious amounts of stale information enabled analysts to understand U-boat habits better and thus direct aircraft with greater skill.

Arguing by analogy to the German use of decryption against Allied convoys, one could estimate that Enigma information, when available, doubled the sweep rate of Allied aircraft in the Bay by halving the area they were obliged to search. This estimate reflects that of the Operations Evaluation Group in their calculation of German benefit derived from decryption of Allied messages compromising the location of convoys,[60] as well as Dönitz's own impressionistic assessment that decryption results were worth an additional 50 U-boats (that is, a doubling of strength) to him.[61]

3

Quantitative Analysis of Action in the Bay

The mere formulation of this "density" theory in precise terms had practical consequences. It showed that the fundamental measure of effective flying is the "miles flown in the operational area," and not, for instance, the hours flown. This point was not always clearly realized, and even at quite a late stage in the war, O.R.S. [the Operational Research Section] sometimes had to criticize arguments which tried to specify desired performances in terms of flying hours, for instance in the drafting of operational requirements for new types of aircraft. Officers who had been tempted into speaking of hours flown, and were then called to order, sometimes tried to justify themselves by the argument that, since the U-boats were assumed to surface at random, they were just as likely to appear at any point of the ocean surface; and that therefore one could theoretically hunt them just as well by waiting over one spot in a Blimp as by flying round; so it was, they argued, the hours and not the distance which is important. This would, of course, be true enough if all one were interested in were to see a U-boat break surface; it is not true if one wants to catch them during the period in which they remain on the surface.

—C.H. Waddington, 1946[62]

To assess the effectiveness of the measures and counter-measures in the Bay of Biscay we need numerical as well as qualitative data.

Basic Bay Data

To treat the problem quantitatively, we need to know the success of the Allied effort, measured in number of sightings; the amount of the effort, measured in number of hours spent on station by patrol aircraft; and the ease of the task, measured in the average number of U-boats present.

Normalizing Competing Factors to Measure Search Effectiveness

The table (table 3) and graph (figure 4) of flying hours, U-boats in the Bay, and U-boat sightings show the results, but not the quality, of the

Table 3.—Data for the Bay of Biscay Offensive[63]

	Sightings	Flying Hours	U-Boat Transits (Allied, Dönitz)		Average U-Boats Present (Allied, Dönitz)		Sunk
Jan 1942	5	350	na	50	9	3.94	0
Feb	3	500	na	51	5.5	4.50	0
Mar	4	400	na	56	6.5	4.65	0
Apr	10	800	na	55	7	5.13	0
May	16	1000	na	49	6.5	4.61	0
Jun	26	2600	50	63	3	5.23	0
Jul	20	3750	65	47	4	4.03	2
Aug	37	3200	80	65	5.5	5.35	0
Sep	39	4100	90	70	6	5.80	1
Oct	18	4100	95	87	7	9.39	1
Nov	19	4600	140	92	9	8.87	0
Dec	14	3400	130	108	8	9.06	0
Jan 1943	10	3130	105	87	7	6.77	0
Feb	32	4400	100	96	6	10.04	1
Mar	42	4600	135	119	8	11.45	1
Apr	52	4200	115	117	7	10.33	1
May	98	5350	120	112	8	11.35	7
Jun	60	5900	57	72	4	7.30	4
Jul	81	8700	78	77	5	10.29	13
Aug	7	7000	na	43	1.5	5.52	5
Sep	21	8000	na	62	3.5	9.10	2
Oct	12	6000	na	75	4	10.33	0
Nov	7	7000	na	44	2.5	8.30	2
Dec	15	6000	na	56	3.5	8.93	1
Jan 1944	22	5000	na	97	4	11.32	3

Notes:
  na = not available.
  May 1943 sinkings include one Italian submarine.

Figure 4
Basic Bay Data
From Morse and Kimball

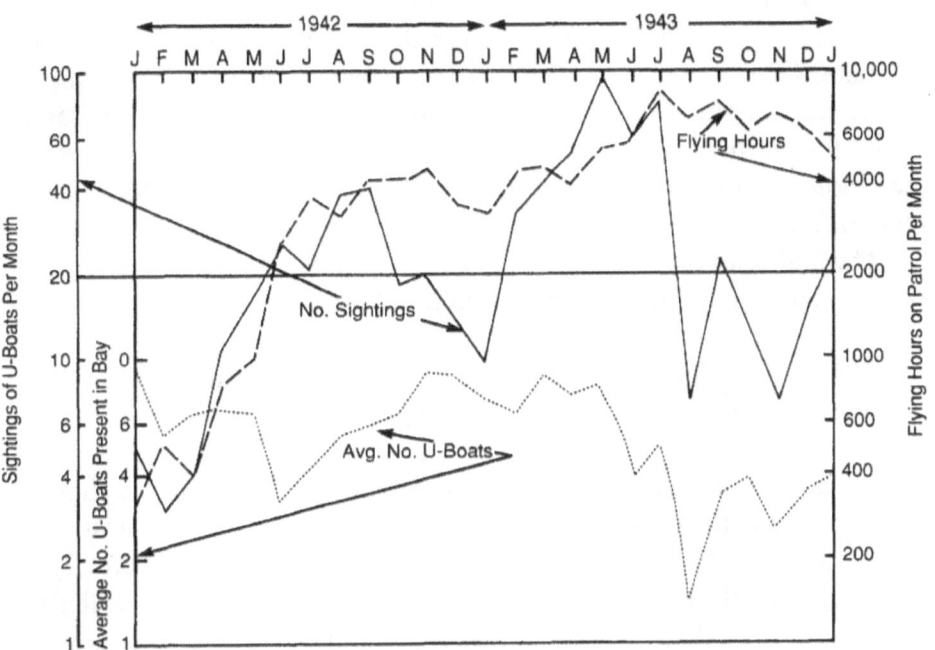

offensive search. The number of U-boats sighted, "bottom line" though it may be, reflects exogenous seasonal factors and the abundance of aircraft and submarines as well as the effectiveness of the former in detecting the latter. For example, in July 1942 20 U-boats were sighted—more than in any previous month of the campaign except for June—but this encouraging fact resulted more from great effort, the large number of U-boats in transit, and the good summer weather than from any great improvement in search performance.

Wartime operations researchers recognized the need to separate the quality of search effort from the quantity and results. Accordingly, they devised the "operational sweep rate" to adjust raw numbers of sightings to compensate for changes in search effort and abundance of targets. The "operational sweep rate" is not "operational" in the military sense of the word, but in the scientific sense: defined in terms of performable acts and measurable quantities.

In the present instance, the operational sweep rate is the number of U-boats sighted divided by the effort, measured in flying hours, and the density of U-boats, measured in boats per square mile. Note that to calculate the operational sweep rate requires an *a priori* estimate of the number of targets in the region. In the Bay of Biscay offensive, interception (even without decryption) of U-boat radio traffic could provide such an estimate. After the war, captured German documents provided definitive information about the movements of U-boats throughout the war. Table 3 shows the wartime Allied estimates (which were remarkably accurate, on the whole) and the more definitive numbers gleaned from Dönitz's *War Diary* after the war.

The operational sweep rate is so called because it measures submarine sightings in square miles per hour and thus has the same units as any other

$$\frac{\text{Submarines sighted}}{(\text{submarines/square mile}) \bullet \text{flying hours}} \qquad \frac{\text{Square miles}}{\text{hour}}$$

sweep rate. The interpretation of this mathematical fact is that one might as well think of the search effort as perfectly examining that many square miles of ocean every hour, instead of (as is actually the case) imperfectly examining more square miles. Knowing that the Bay comprised 130,000 square miles of searchable area, we may compute for each month the operational sweep rate of the offensive forces in the Bay of Biscay campaign. (See figure 5.) Since it is the only sweep rate of interest to us, we will often refer to the operational sweep rate simply as "the sweep rate."

The operational sweep rate derived from Allied data differs from that derived from the more authoritative Dönitz data primarily in that the former shows huge spikes in January and May of 1942 and portrays a lower search effectiveness during most of the time radar was in use. Reviewing the data, we can see that while the Allied side had a roughly accurate estimate of the number of U-boat transits in most months, they underestimated the average density of boats in the Bay by overestimating the speed with which boats crossed the Bay.

Sternhell and Thorndike's data,[64] shown in table 3, include the number of transits for each month from June 1942 through July 1943; comparison with figure 4 (Morse and Kimball's graph[65]), which includes the average number of submarines present in the Bay for each month, leads to the conclusion that the submarines took a little more than 40 hours to cross the Bay. According to Sternhell and Thorndike, a 42-hour passage entailed 21 hours on the surface. But Sternhell and Thorndike also present a detailed calculation based on 13 surfaced hours per transit, which they indicate elsewhere corresponds to a 125-hour passage.[66] These points of confusion, as well as the fact that the Allied data represent intelligence estimates and the German data represent straight recordkeeping, suggest that we should use the German figures.

Turning to the early 1942 spike, Sternhell and Thorndike, writing before knowledge of Enigma decryption became public, mention that radio direction-finding improved the antisubmarine sweep rate of surface ships

Figure 5
Bay of Biscay Operational Sweep Rate
From Allies' and Dönitz's Data

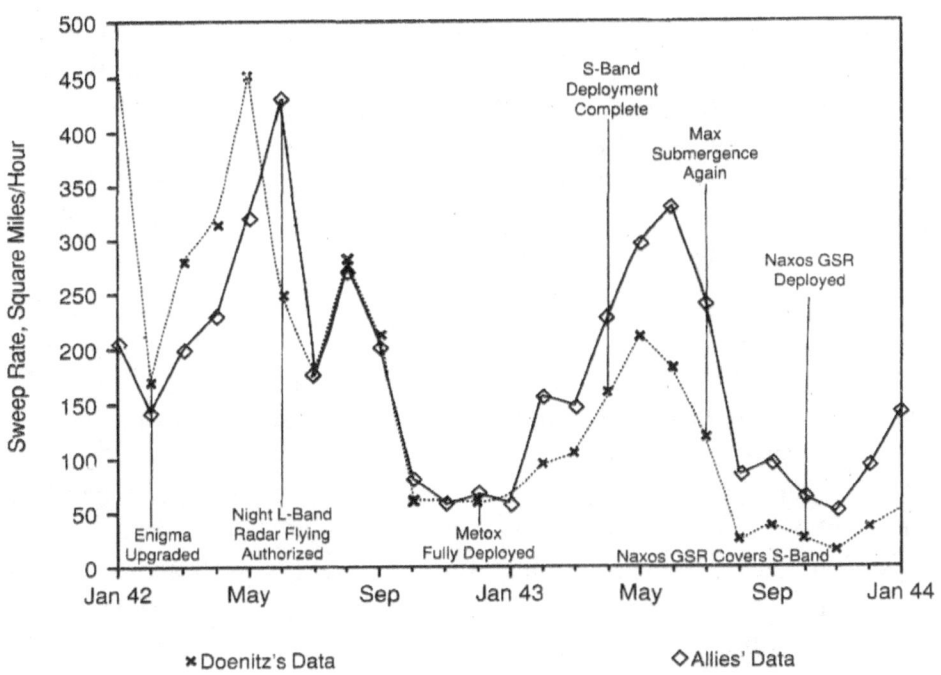

by a factor of three or four.[67] A similar advantage conferred by decryption may explain the extraordinarily high sweep rate of January 1942, when Allied aircraft did as well without radar and the Leigh Light as they ever did with them, and did about three times better than they were to do the next month with the same equipment. January was the only month in 1942 in which the British could decipher U-boat Enigma messages consistently; addition of a fourth wheel to the Enigma machine in February rendered decryption impossible for the rest of the year.[68] I would offer that, to keep the operations researchers from realizing that some source of information such as Enigma decryption must exist, the Allied figures were "cooked," overstating the abundance of U-boats in the Bay that month to make the number sighted by ostensibly undirected search less surprising. In a parallel incident, American operations researchers received true statistics from the hunt for U-boats off the East Coast of the United States and correctly deduced that the ostensible means of knowing where to look—radio direction-finding—was a cover for some much more powerful method.[69]

By the last quarter of 1943, the British could, with little or no delay, decipher both the U-boat version of Enigma and the Home Waters version, used by U-boats in the Bay of Biscay.[70] Exploitation of such information undoubtedly contributed to the growing effectiveness of antisubmarine patrol aircraft in the closing months of 1943.

Successful exploitation of decryption could cause the sweep rate to transcend the limits of equipment performance, but it cannot explain 1943's consistent high rate based on Allied data compared with that based on Dönitz's data. (Nor can deliberate Allied fudging of wartime data to conceal Enigma decryption, because this time the discrepancy is in the wrong direction.)

Seasonal Adjustment by Fourier Analysis and Regression

The graphs of the operational sweep rate in figure 5 show peaks in the summers of the two years. Morse and Kimball do not mention any possible impact of the seasons on search effectiveness, yet seasonal change in the amount of daylight would certainly be expected to affect visual sighting of submarines. Sternhell and Thorndike, in their analysis of the Bay operations, repeatedly cite seasonal change as a possible explanatory variable,[71] but do not attempt to quantify or compensate for the effects of seasonal variation. Both pairs of authors concentrate on the effects of changes in equipment and tactics.

Of course, the effects of changes in equipment and tactics are the subject of this book as well, but we ought first to see if any effects remain after correction for seasonal change.

Fourier analysis—also called "spectral analysis" because of its prism-like separation by frequency—provides one way to correct for seasonal change. Commonly used in signal processing, such analysis capitalizes on the remarkable fact that any continuous function has a unique decomposition into sinusoidal waves. In other words, Fourier analysis sorts out a complicated waveform into a (possibly infinite) set of simple sinusoidal waves that—together with a constant equal to the mean observation—add up to the original curve. The decomposition's uniqueness is important because it precludes the possibility that some other set of sinusoidal waves adds up to the same function.

Often, as in the case at hand, the data provided are not a continuous curve but merely a set of evenly spaced points. In such cases, Fourier analysis provides an average value and a unique finite set of sinusoidal waves whose sum passes through all the given points and whose shortest period is twice the spacing of the points.

Twenty-four observations (the algorithm requires an even number—we discard January 1942 because of its extraordinary Enigma success) therefore uniquely determine 12 sinusoidal waves and a mean. We can view the result in two ways: a time-domain graph (figure 6) showing the original signal and its component waves, and a frequency-domain graph (figure 7) showing the amplitudes of the different waves. In the latter, the waves are numbered according to how many cycles they complete in two years. In the former, the mean and the four longest cycles appear.

Fourier analysis of the sweep rate, considered as a waveform, reveals a strong 12-month "hum." As one would expect, this cycle peaks midyear, albeit in May rather than June—perhaps because some annual effect other than daylight, such as weather, exerts influence upon the sweep rate. Extracting this annual cycle as one might filter an offending tone out of an audio signal, we obtain a graph of the seasonally adjusted sweep rate. (See figure 8.) Introduction of new equipment by the Allies and Germans accounts for peaks and troughs, respectively, as it should. Unfortunately, this method discards all annual periodicities, not just those caused by the seasons, so we may also be discarding some of the effects we are trying to observe if they just happen to fall into a summer-centered annual cycle. In fact, introduction of British equipment tended to occur in the spring and German countermeasures tended to appear in the fall, so we are almost certainly underestimating their effectiveness if we discard the whole annual swing of search effectiveness.

Adding another assumption allows use of a more familiar statistical method, linear regression. The needed assumption is that seasonal dependence stems from the annual fluctuation in the daylight. Linear regression using the sinusoidally varying daylight as an independent variable will again result in a sinusoidal curve of "explained" search rate, which can be subtracted from the original to yield a residual, seasonally adjusted search rate. This approach results in the following equation for the search rate:

$$\text{search rate} = 16 \cdot \text{hours of daylight} - 42.$$

Figure 6
Fourier Analysis: Longest Cycles

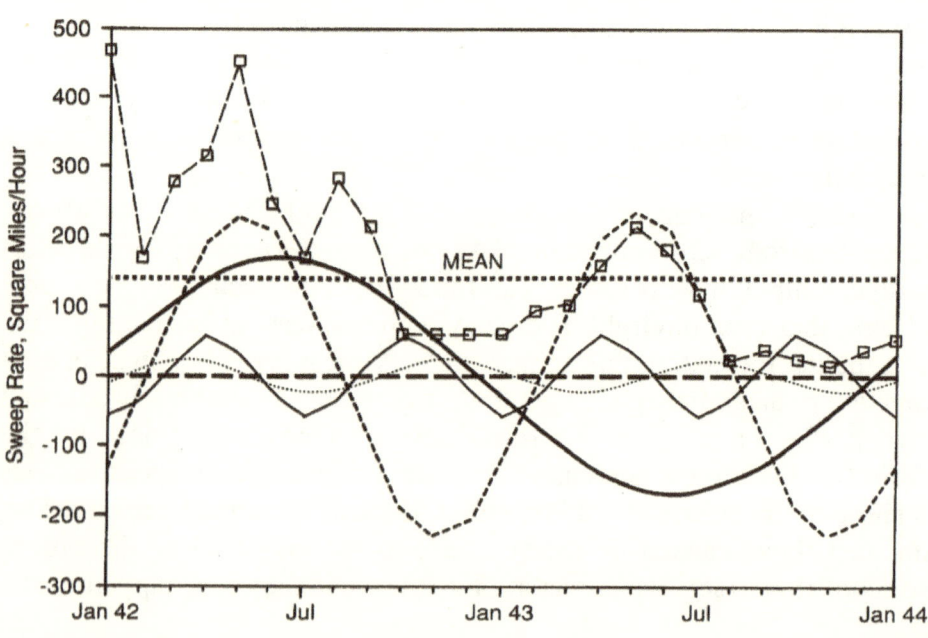

Figure 7
Spectral Analysis of Bay Sweep Rate
Amplitudes of Fourier Components

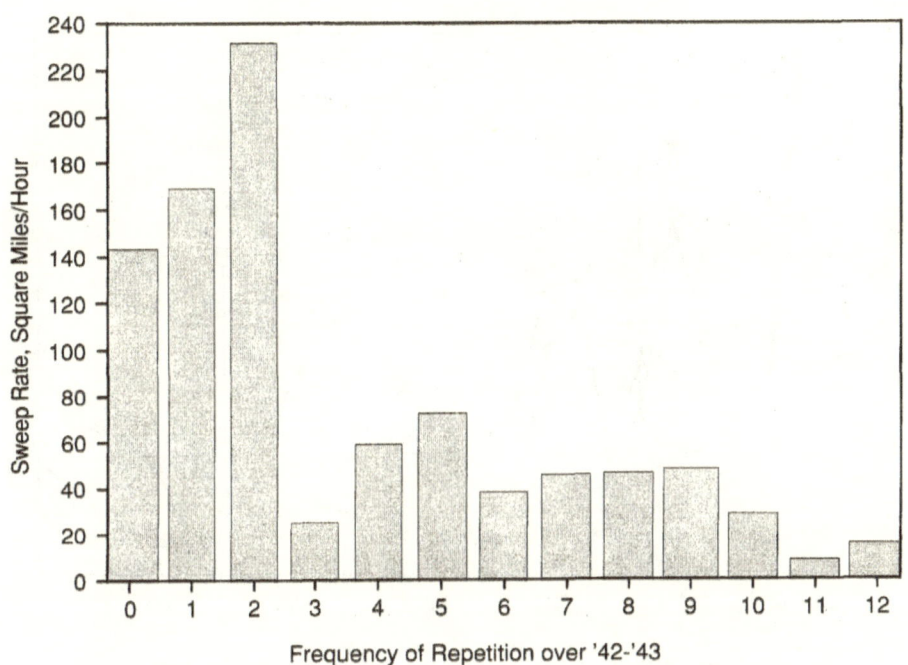

The residuals—the observed sweep rates minus those returned by the bilinear regression—make a seasonally adjusted rate that corresponds closely to that found using the Fourier analysis method. (See figures 8 and 9.) Unlike Fourier analysis, however, a linear regression's goodness of fit can be assessed quantitatively: the idea that different rates apply by day and by night accounts for 14 percent of the variance in sweep rate. This low value suggests that something other than daylight hours determines sweep rate. An even more compelling objection to the regression fit is that we can assign no meaningful interpretation to the negative constant term. Forcing it to be zero on the grounds that a negative sweep rate is nonsense only worsens the fit.

Seasonal Effects Do Not Fully Explain Fluctuations

The fluctuations remaining after seasonal adjustment show that some effect other than the variation in the length of day is at work. Bad weather comes to the Bay at the same time as short days, but it reduces visibility less

Brian McCue

Figure 8
Seasonally Adjusted Bay Sweep Rate
From Dönitz Data

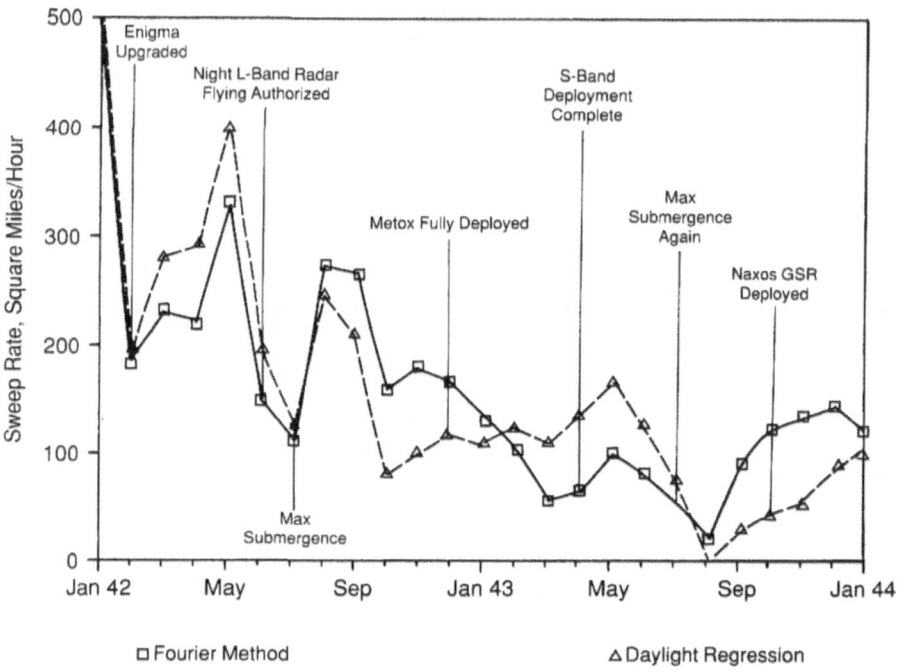

than one might expect: median visibility in the Bay peaks in June at 18 nautical miles and reaches a minimum of 10 nautical miles in the winter. An observer, as we will see later, is extremely unlikely to sight a submarine at a range of 10 or 18 miles in any case. Another telling statistic is the number of days per month with visibility less than five miles. One can expect such conditions about 15 percent of the time except in late summer and early fall, when visibility, as measured in this sense, improves.[72]

Even with all the "12-month hum" removed, considerable variation remains. Significantly, the fluctuations coincide with the combatants' attempts to influence the operational sweep rate in one direction or the other: the graph of seasonally adjusted operational sweep rate generally shows improvement upon the introduction of new British equipment and decline after the introduction of new German countermeasures. This graph proves that changes in equipment and tactics, and not merely the passage of the seasons, affected the operational sweep rate of Allied patrol aircraft.

Byproducts of Spectral Analysis

Human affairs often appear to display periodicity, sometimes attributed to the oscillation of a system controlled by feedback; thus we have the notion of a "business cycle," a "decision-making cycle," or, as in the present case, an "RDT&E [research, development, test, and evaluation] cycle."[73] Frequent references in the literature[74] to the "orientation-observation-decision-action cycle" and the desirability of "getting inside the adversary's decision loop" suggest acceptance of the notion that such cycles exist in military tactics. Others have discerned far longer cycles in history, including one of particular interest here, a 500-year cycle in the balance between European maritime traders and the raiders who would prey upon them.[75] Let us take for granted this received view that there exist cycles and try to find them in the present case. We may attempt to discover British and German periodicities in the introduction of measures

Figure 9
Second Approach to Seasonal Adjustment
Linear Regression on Daylight Hours

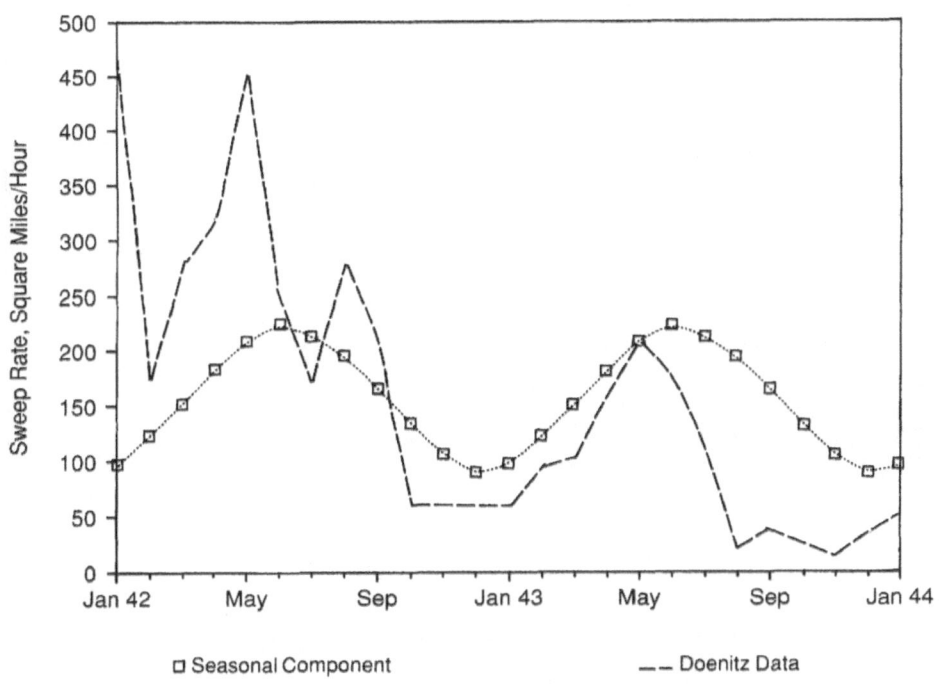

and countermeasures by attaching either side's milestones to the waves found in the spectral analysis. We would expect British innovations to appear when the British cycle, if any, bottoms out: i.e., when search effectiveness is at its lowest and about to rise. Conversely, a German cycle would show the German milestones clustered in areas where British search effectiveness is at its highest and about to decline.

Applying the British and German milestones to the various waves found by spectral analysis creates the best fits with the 8-and 4-month cycles (that is, those that repeat three and four times in the 2 years) respectively. (See figures 10 and 11.) As the bar graph of amplitudes shows, these are not the strongest components of the sweep rate "waveform," but the milestones relate to them roughly as described above. The deployment of Metox is an exception in the German cycle, but the reader will recall that the Metox receiver was an expedient countermeasure, pressing a preexisting French receiver into use against the L-band ASV Mark II almost as soon as nighttime use of that radar began: RDT&E had already happened under French auspices.

Figure 10
German-Looking Cycle
Sternhell and Thorndike Milestones

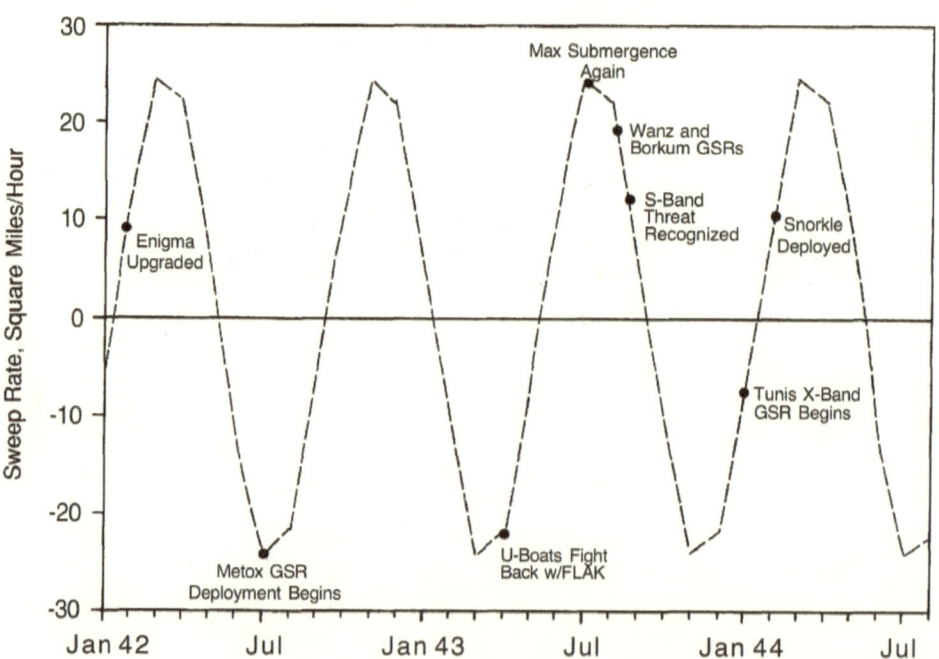

Figure 11
British-Looking Cycle
Sternhell and Thorndike Milestones

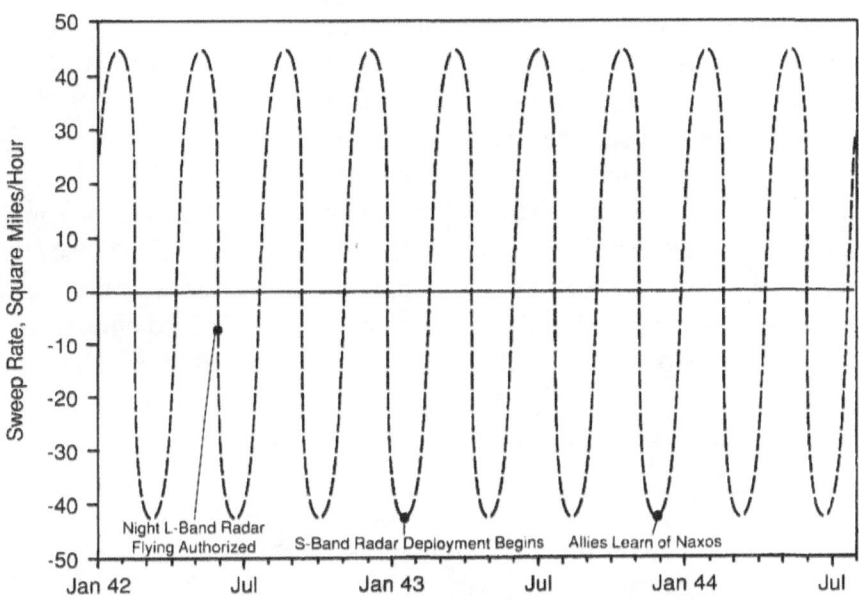

The British cycle, as elicited from the data by this method, has greater amplitude and shorter period than the German cycle, suggesting superior speed and efficacy of innovation. These findings certainly seem reasonable in light of the conflict's final outcome. The evidence for the cycles' existence, though tenuous (so tenuous that no further mention of them will be made here), constitutes a stronger affirmation of their existence than is presented in typical explications of the received view.

Autocorrelation Analysis

Autocorrelation analysis, a statistical approach closely related to spectral analysis, measures the statistical correlation of elements of a time series with their predecessors.[76] The analysis results in a series of correlation coefficients, the first being the correlation of all elements with their immediate successors, the second being the correlation of all elements with their successors' successors, and so on. After seasonal adjustment, we may apply autocorrelation to investigate the average length of what the combatants might have considered "winning streaks."

   The analysis produces a bar graph of correlation coefficients: in
the present case the ith coefficient shows the correlation of seasonally
adjusted sweep rates i months apart. (See figure 12.) The first four bars are
positive and fairly large, whereas the next eight are small or even negative,
seemingly indicating that—from the standpoint of a participant—one
may plan four months ahead on the basis of current seasonally adjusted
experience, but that thereafter prediction is impossible.
   As with all statistical estimates, however, one must consider the results of
the autocorrelation analysis in light of the fact that even a totally random series
of numbers could, through chance alone, display some pattern: measures of
statistical significance weed out patterns so weak as to resemble those arising
from happenstance. In autocorrelation, apparent correlations of widely spaced
terms must be considered in light of the already-determined correlations at
lesser spacings. For example, if successive terms strongly correlate, some
level of correlation of alternate) terms is inevitable; any seeming correlation
of alternate terms is significant only if it exceeds that level.

Figure 12
Autocorrelation of Sweep Rate
Seasonally Adjusted Rate from BdU Data

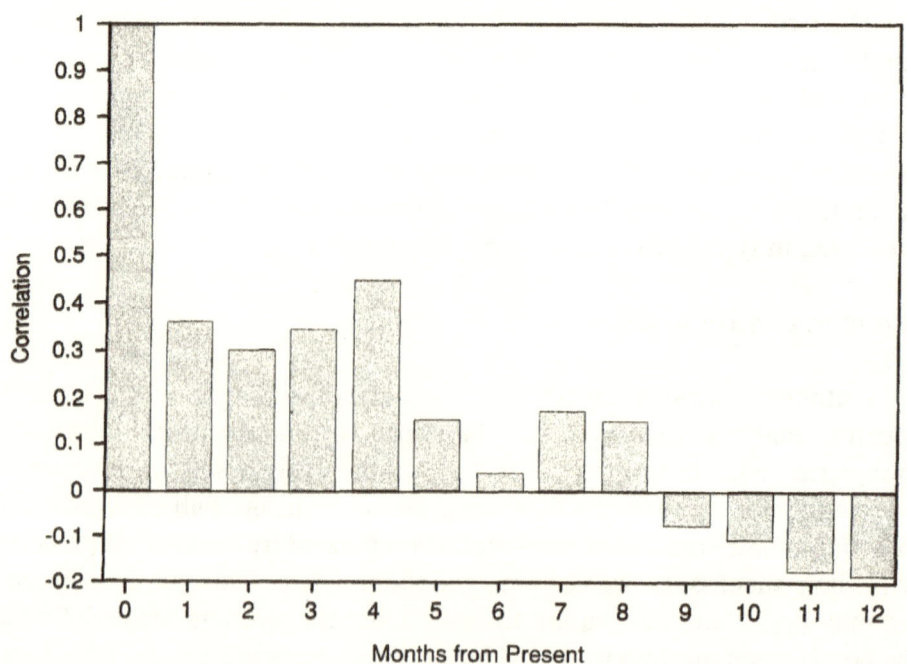

In the case of the seasonally adjusted operational sweep rates, the correlations at lags of 1, 2, 3, and 4 months just barely fail to meet the usual test of significance, and the succeeding terms fail the test by wide margins. In short, all the terms, according to this test, might as well be zero. It is worth noting, however, that some contention surrounds the criteria for significance in such tests.[77] Given the narrow margin by which the first four months were rejected, we may conclude—albeit with diminished confidence—that innovations, of either equipment or tactics, tended to occur every five months or so, radically altering the contest.

The Response of the Sweep Rate to Innovation

Turning back to the original data, one can see a distinct pattern of high effectiveness when new Allied equipment appeared, followed by a period of lower effectiveness. Sweep rate leapt to a record level at the start of night L-band flying, and plummeted the next month when the first few Metox units entered service. Sweep rate rose dramatically when the S-band ASV Mark III first came into use, then dropped and rose slowly to the level at which Dönitz became alarmed and introduced a variety of behavioral changes—as well as the ineffective Wanz and Borkum receivers—whereupon sweep rate dropped to an all-time low, then rose steadily without the introduction of new British equipment.

In general, then, the dynamics of the Bay search process include negative feedback: any change triggers a backlash. Simple analogies would be the rebound of an automobile suspension back to—and perhaps past—its original position after the car hits a bump, or the vibration of a plucked violin string. A more complicated analogy would be the "price corrections" seen in commodities markets. In each of these cases, a *restoring force* (the car's springs) acts to counter any perturbation, and *damping* (the car's shock absorbers) slows the system's response to any force. When perturbed, the system responds to the restoring force and—unless the damping is very strong—overshoots its original position somewhat. After an overshoot, the system oscillates back and forth in ever-smaller swings, settling down asymptotically. The asymptotically reached final position need not be the original one: for example, the car may have bounced in response to a heavy load dropped into the trunk, or the price of the commodity may have gyrated in response to a shortage.

In the Bay search operations, the introduction of new equipment clearly displaces the operational sweep rate to a new value. Once displaced, the sweep rate tends to move back towards its original position. Such motion could result from the introduction of a countermeasure by the other side,* or from a variety of human factors: a reactive change of tactics by the other side; overconfi-dence in the new equipment, once its efficacy has been established; or the Hawthorne effect, under which workers initially respond favorably to any change in the work environment, whether it is an improvement or not. Then the novelty of the change wears off; the fact that Dönitz had to issue a second maximum-submergence edict even though he had never rescinded the first shows that tactics could simply wear off over time, as commanders stopped adhering to doctrine.

This study does not aim to distinguish among these human factors, but only to show the existence and character of the oscillations. Physical oscillators can be characterized in terms of their response to

Figure 13
Innovations in Bay Search

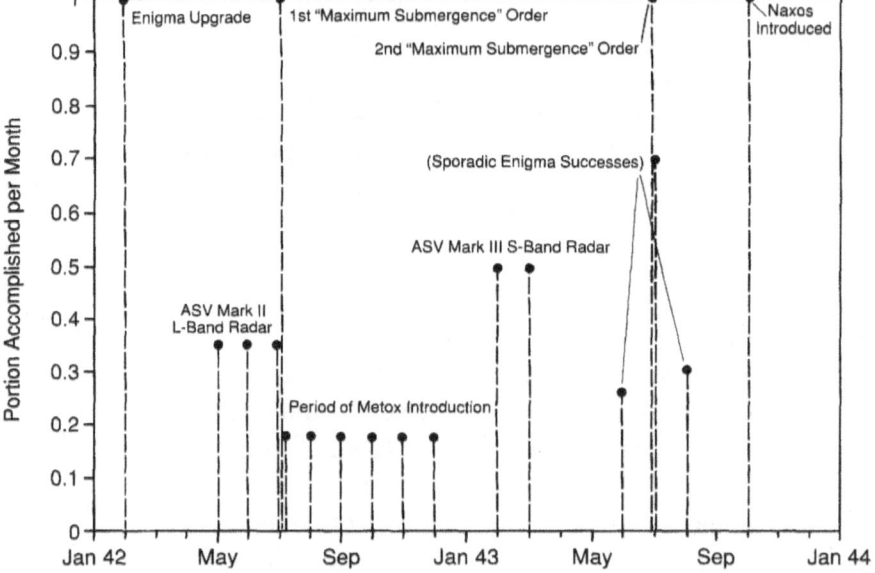

---

*   But NOT the introduction of a second innovation by the side that introduced the first. The whole point of the present analysis is that it extracts from the data the net effect of a single innovation, separate from the effects of later innovations by the same side.

shock—instantaneous displacement and immediate release. For example, a mechanic can assess the "ride" of a car—its response to bumps in the road—by applying a single downward push to the fender: the car's response to this shock indicates how well it will filter vibrations imparted by the road. Similarly, a musician can assess the acoustics of a concert hall by listening to the echo of a single handclap. Indeed, the response of such a system to any "forcing function" can be typified by its response to a shock—its "impulse response."

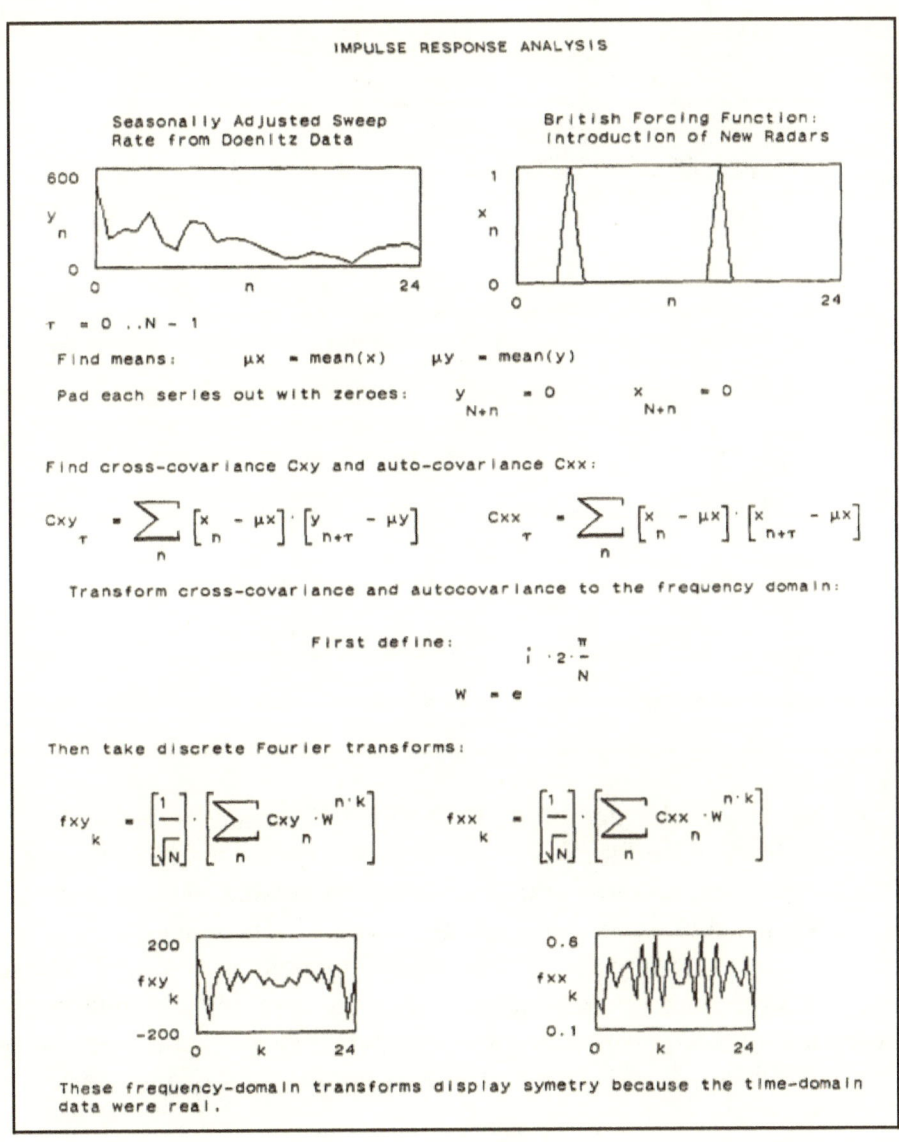

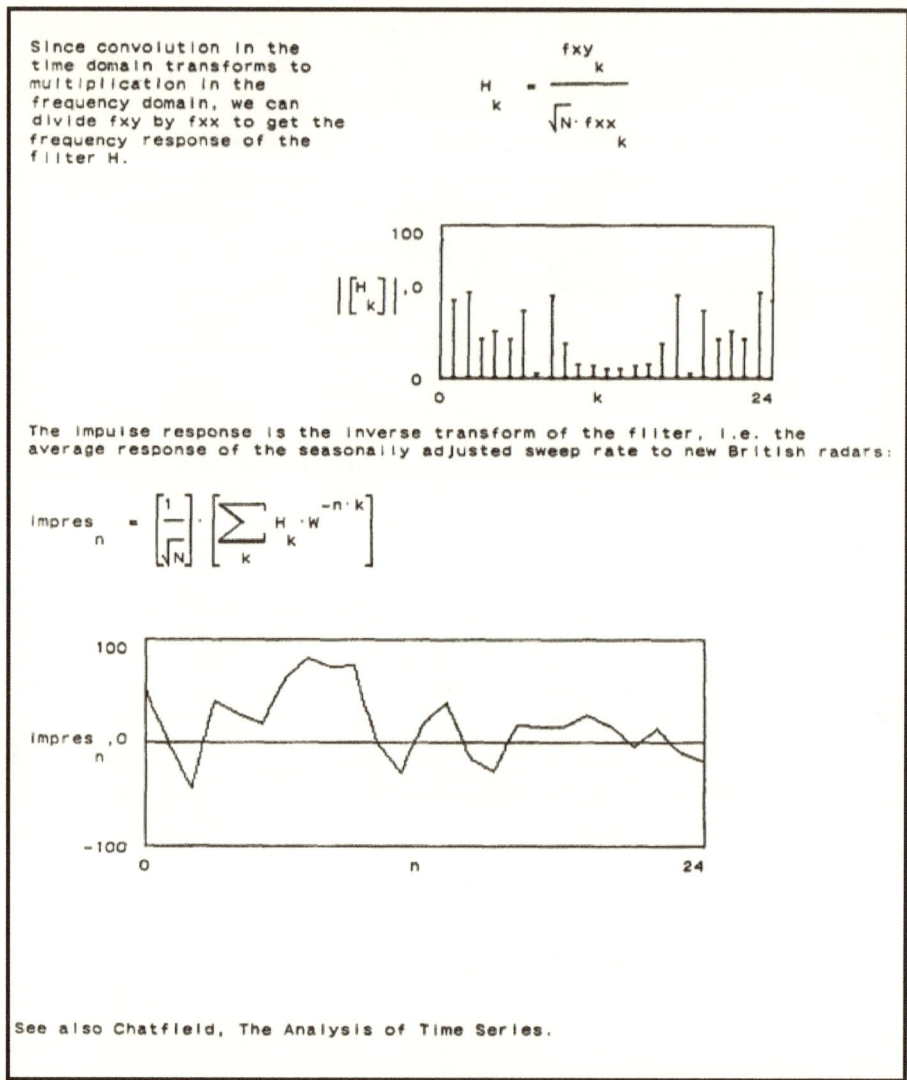

Since convolution in the
time domain transforms to
multiplication in the
frequency domain, we can
divide fxy by fxx to get the
frequency response of the
filter H.

$$H_k = \frac{fxy_k}{\sqrt{N} \cdot fxx_k}$$

The impulse response is the inverse transform of the filter, i.e. the
average response of the seasonally adjusted sweep rate to new British radars:

$$Impres_n = \left[\frac{1}{\sqrt{N}}\right] \cdot \left[\sum_k H_k \cdot W^{-n \cdot k}\right]$$

See also Chatfield, The Analysis of Time Series.

We have no trouble determining the forcing function—innovation—in Bay operations. (See figure 13.) Except for the Enigma innovations in 1943 (when German radio traffic became unreadable, but for less than a full month), each innovation receives an arbitrary value of unity: an innovation is an innovation. Those implemented in a single month, such as maximum submergence, receive a value of unity in that month. Those that took some months to implement, such as the installation of the ASV Mark II, are prorated: the 3-month onset of the Mark II is counted as a third of an innovation in each of the months. The Wanz and Borkum receivers, doomed to failure because they listened on the wrong wavelengths, do not appear. The large May 1942 increase in the sweep rate derived from the Dönitz data (figure 5) justifies crediting the ASV Mark II L-band radar with introduction in that month rather than the next one. (As the reader will recall, the nighttime use of this radar was made possible by the Leigh Light, an expedient measure developed in the field in violation of orders. Such measures might well see use before their ostensible introduction. Indeed, one authoritative source cites Leigh Light flying as beginning in May.[78])

We may consider the operational sweep rate to be a system perturbed by either of two forcing functions, British innovation (the ASV Mark II and ASV Mark III radars) or German innovation (the Metox and Naxos receivers as well as the periods of Enigma unreadability and of maximum submergence). Using Fourier methods to relate the seasonally adjusted operational sweep rate to these forcing functions produces the sweep rate's impulse response: its reaction to a single innovation. The graphs of sweep-rate response to innovation (see figures 14 and 15) show several things: correct initial movement followed by oscillation, British predominance, and a stronger response to a change of tactics than to a change of equipment.

Computed Responses Look Plausible

Upon their introduction (that is, during month 0 as shown on figures 14 and 15), the various measures and countermeasures work as intended: new radars raise the sweep rate, German countermeasures lower it. This observation provides "face validity" to this method of discerning responses to innovation.

The responses cross the abcissa—indicating negation—somewhere between one and two months after their introduction. They then behave like damped oscillators, zig-zagging toward zero change in a series of narrowing swings: a restoring force acts on sweep rates, even apart from seasonal effects.

For example, after the low point following full introduction of Metox, the seasonally adjusted sweep rate (see figure 9) begins to creep up even before the point at which S-band deployment begins. After the low point created by the various reactions to the S-band ASV III, Allied

Figure 14
Response to British Innovations
Oscillation After Innovation "Shock"

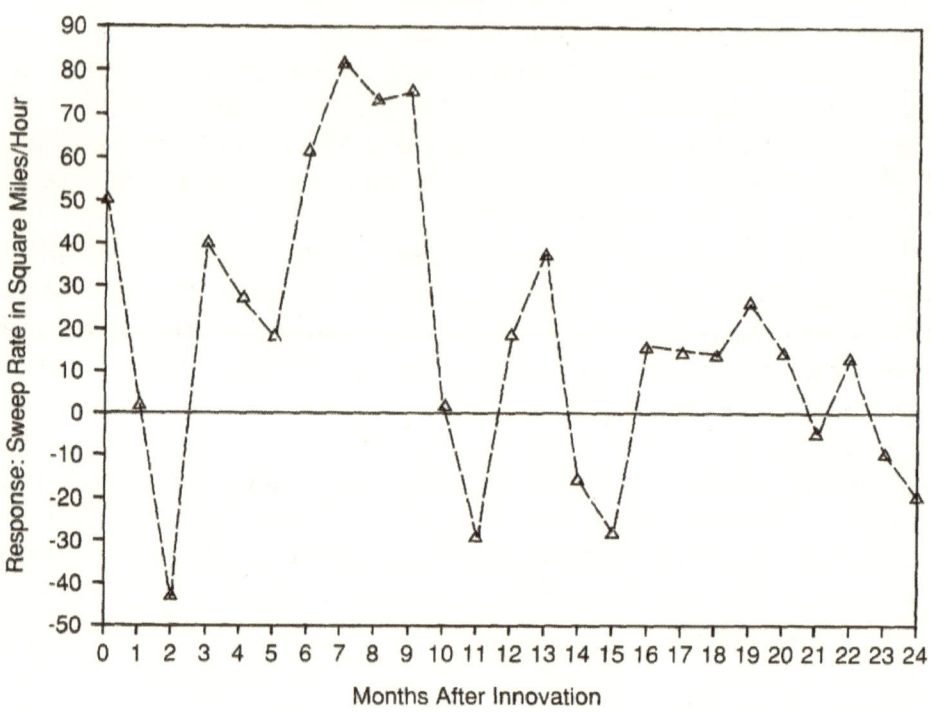

effectiveness again increased steadily, long before the introduction of the X-band radar. The damping of the seasonally adjusted sweep rate's impulse response to both British and German innovations shows this point mathematically.

Some arms-race theorists might think the sweep rate's oscillations confirm a theory that the other side will soon vitiate any improvement, or even that the parties to an arms race are really racing with themselves, not each other.[79] I do not subscribe to this moral at all! The graph of the seasonally adjusted rates shows that the sweep rate almost always exceeded the 125-square mile/hour rate that prevailed before the introduction of night flying with radar. Also—as we shall see later—sweep rate measures only the effectiveness of the airplanes in finding submarines; the period of retrenchment in 1943 cost Dönitz many U-boat at-sea days and undoubtedly saved some merchant ships.

Responses Favor the British

Both response curves favor the positive side of the swing—the region that favors the British. The sweep rate's impulse responses to the two sides' innovations appear to show that the British innovated better as well as faster: the responses are predominantly positive, indicating an increased sweep width.

One may well ask, then, how it is that the seasonally adjusted sweep rate shows an overall trend downward when the British innovations worked better than the German ones. The answer lies in the maximum-submergence tactic, an effective German counter-measure that was introduced twice during the period under study.

Figure 15
Response to German Innovations
Oscillation After Innovation "Shock"

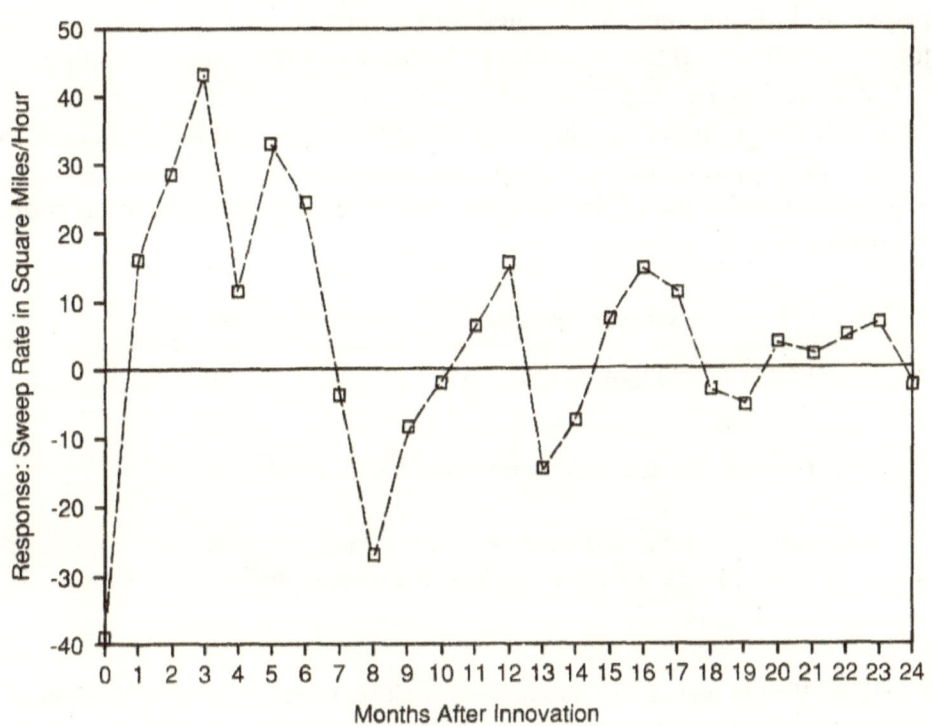

Maximum Submergence Most Effective German Innovation

The impulse response to the maximum-submergence counter-measure, computed separately, displays much less vitiation than the response to German innovations overall. (See figure 16.) Early operations researchers emphasized that their role was not to specify the creation of new gadgets but to research ways of exploiting equipment to the utmost.[80] The long-lasting benefits of maximum submergence confirm the view that one need not improve equipment to improve results.

How Much Was Enough?

To further the ultimate end of saving merchant shipping, the Allies had not merely to sight submarines in the Bay, but to sink them. Selecting, from a boat-by-boat tabulation of the fates of German submarines,[81] those sunk by aircraft in the Bay of Biscay, we may form the rightmost column of the table of basic Bay data (table 3).* The efficacy of the flying effort with respect to kills can then be adjusted for the amount of flying effort and of target density as before, resulting in a *kill sweep rate* entirely analogous to the operational sweep rate but hinging upon the number of submarines sunk, not the number sighted. Again, this measure of effectiveness is expressed in square miles per hour, and may be interpreted as the number of square miles' worth of the submarine-infested Bay that an aircraft could exterminate in an hour:

$$\frac{\text{Submarines sunk}}{(\text{submarines/square mile}) \bullet \text{flying hours}} = \frac{\text{Square miles}}{\text{hour}}$$

Wartime operations researchers compiled such a statistic.

Alternatively, one could leave the sweep-rate definition as it is and form a second statistic, the probability of kill given sighting. (See figure 17.)

---

*   This effort reveals a few submarines sunk by mines in the Bay. Minefield-induced channelization may have reduced the area in which submarines could safely operate. If patrol aircraft capitalized on this effect, the later sweep rates presented by Morse and Kimball are in error.

Figure 16
Response to Max Submergence
Oscillation After Innovation "Shock"

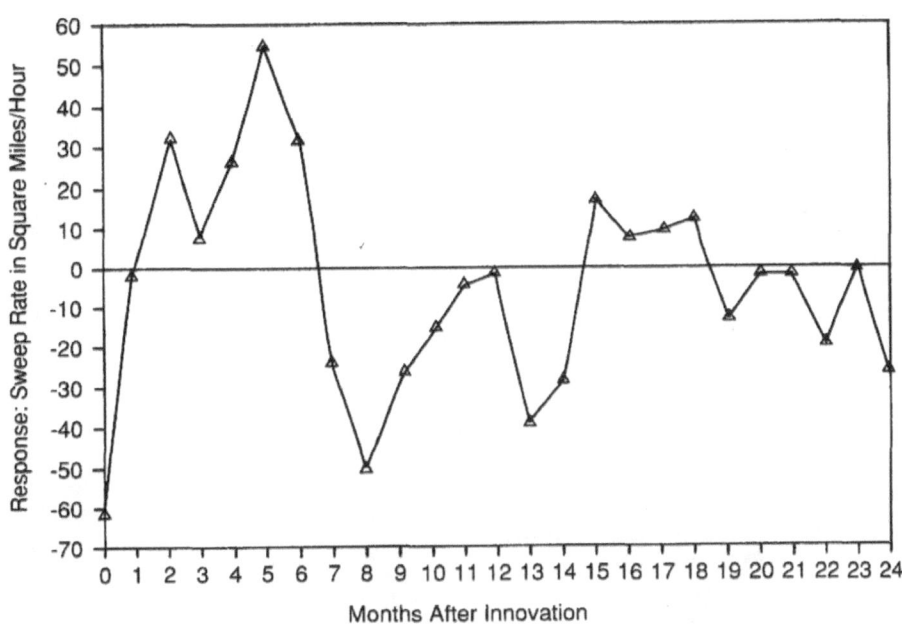

This probability-of-kill-given-sighting, or P(K|S), statistic—smoothed by a five-month moving average—will be useful later because it separates the changing lethality of the weapons and tactics used against the submarines, once sighted, from the effectiveness of the search for them.

Wartime operations researchers convincingly attributed the increasing lethality of the attacks to changes in the attacks themselves, not to the sightings leading up to them.[82] These changes included decreasing the depth to which the depth charges (DCs) sank before exploding, improving the explosives in the charges, abandoning the practice of aiming ahead of the submarine, and increasing the spacing between the charges. Thus it is not surprising that the probability of kill given sighting and the operational sweep-rate display little correlation; even less correlation if one compares the seasonally adjusted versions of each measure obtained by the Fourier method. We will implicitly rely on this fact later, when we ask "What if . . . ?" questions: we will use historical values for probability of kill given sighting with alternative sweep-rate values stemming from hypothetical considerations such as "What if the Germans adopted *minimum* submergence tactics in order to cross the Bay faster?"

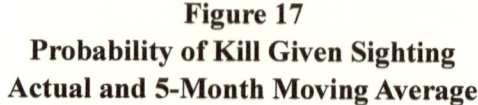

Figure 17
Probability of Kill Given Sighting
Actual and 5-Month Moving Average

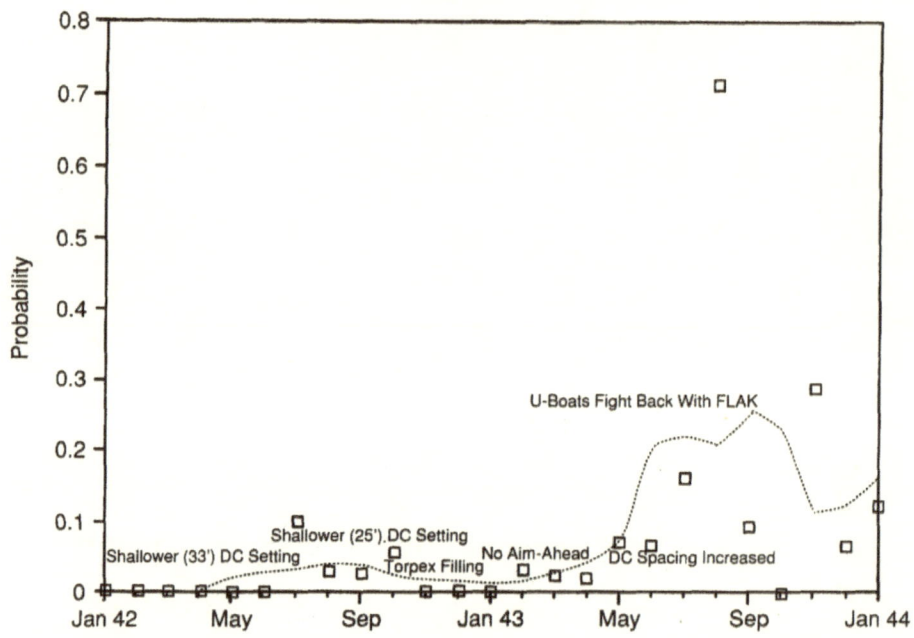

Bay Search Efforts Proved Worthwhile

Unlike strategic bombing of Germany,[83] air patrol of the Bay of Biscay showed a clear "profit." An hour of flight in Bay patrol operations consumed, including maintenance, about 33 man-hours.[84] A total of 104,080 flight hours in 1942 and 1943 thus consumed about 3,400,000 man-hours of effort, not counting airplane construction time—an element justifiably left out on the grounds that few airplanes crashed during search operations. Bay patrol sank 44 U-boats, so about 78,000 man-hours were needed to sink a U-boat. Construction of a U-boat, on the other hand, required between 250,000 and 800,000 man-hours (depending on the type of U-boat[85]), so Allied time spent on Bay patrol clearly paid off. Figure 18 shows submarines sunk per man-hour on a monthly basis for the period of concern/retaining the electronic-warfare milestones for reference. On the scale of this graph, the fraction of a U-boat built per man-hour is not visible: the months in which the Bay patrol sank any U-boats at all result in a man-hour balance sheet strongly in favor of the British.

But balance-sheet profit was not going to win the war for the Allies. The purpose of offensive search in the Bay was to save merchant shipping. The same bombers and aircrews could accomplish the same mission through defensive patrol—air escort of convoys—or, less directly, by bombing U-boat bases. Alternatively, the bombers could go even farther into Europe and bomb the U-boat-building shipyards or other strategic targets. The British recognized these choices at the time, not only as a question of airplane use,[86] but also as a factor in deciding which units should get the S-band radars first. Some made the argument that submarine patrol, offensive or otherwise, was more important than strategic bombing, and also less likely to compromise the S-band radar.[87]

Figure 18
U-Boats Sunk Per Man-Hour of Bay Effort
33 Man-Hours per Flight Hour, All Told

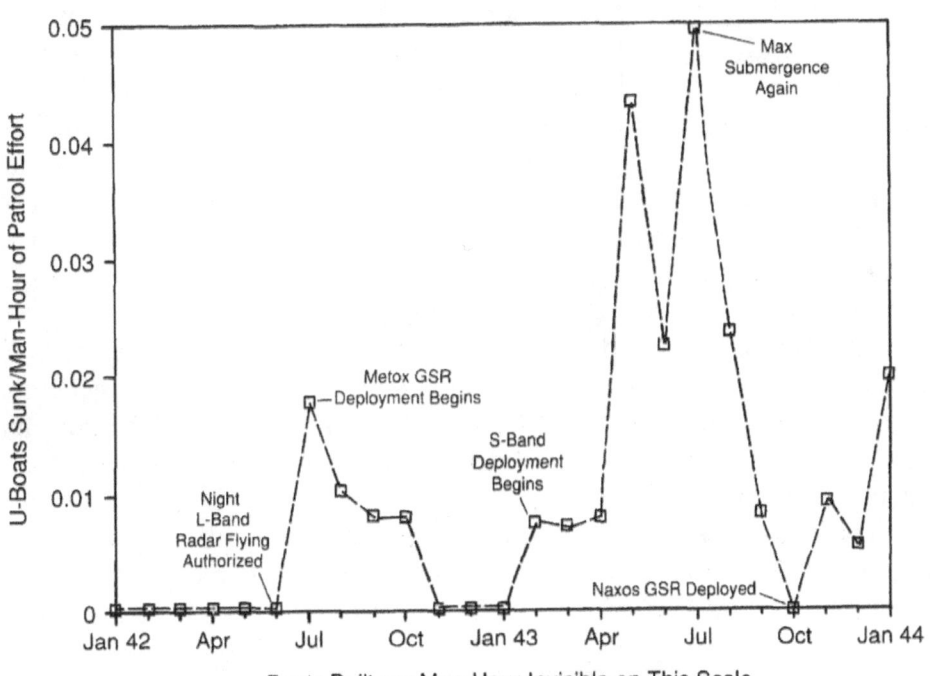

Boats Built per Man-Hour Invisible on This Scale

This argument failed, leading almost immediately to the Luftwaffe's discovery of the "Rotterdam Gerat" as described earlier.

To decide such issues, we (or the wartime allocators of flying effort) need to relate flying hours to U-boats destroyed. Seasonal adjustment is inappropriate in addressing the opportunity cost of searching the Bay; the seasonal changes in search effectiveness need to be left intact.

Morse and Kimball present the following logic in comparing the relative effectiveness of air attacks on U-boat pens, defensive air patrol around convoys, and offensive use of air assets in the Bay of Biscay,[88] all measured by the decline in the number of merchant ships sunk. The period of study is the second half of 1942, when offensive flying in the Bay had become effective because of radar and the Leigh Light. Examination of convoy losses suggests that 100 air sorties—about 1,000 hours—flown in defense of threatened convoys saved about 30 merchant ships. Ignorance of which convoys were in fact threatened diluted the rewards of flying effort by a factor of 10, so that in nonselective escort of convoys three ships were saved per 1,000 hours of flying. Based on photo reconnaissance after air raids, Morse and Kimball estimate that the 1,100 sorties flown against U-boat bases cost the Germans about 15 submarine-months-at-sea in delays, without sinking any submarines at all. Because a submarine sank about eight-tenths of a ship per month on patrol, the bombing saved 12 ships—not even as many as uninformed escort of convoys would have saved. Turning to the Bay operations, a U-boat could expect an operational life of 10 months at sea, so a sunken U-boat was worth eight merchant ships. A hundred sorties would, on average, sink half a U-boat, saving four ships: a better application of flying effort than anything but focused defense of threatened convoys. Morse and Kimball go on to cite the delaying effect of air patrol on submarines in transit, worth about another merchant ship per 100 sorties.

Sternhell and Thorndike[89] cite 1.6 merchant ships sunk per U-boat month at sea in the same period, exactly double the figure given by Morse and Kimball. Morse and Kimball may have mistakenly used a figure for submarine "productivity" per month, not per month at sea—the U-boats spent about half their time at sea.*

The sources are in approximate agreement as to the life expectancy of a U-boat: Morse and Kimball say 10 months, Sternhell and Thorndike say

---

* Postwar data, as we shall see later, indicate a figure close to one ship per month of at-sea U-boat time.

11, and both appear to refer to actual time spent at sea.[90] Using Sternhell and Thorndike's numbers in Morse and Kimball's calculation would leave the effectiveness of convoy protection the same, but would double the result of attacks on U-boat repair areas to 24 merchant ships saved per 1,100 sorties, and would double the result of Bay patrol to 10 merchant ships saved per 100 sorties. The broad conclusions remain unchanged: Bay patrol is the second most effective mission for these aircraft, inferior only to protecting convoys known to be threatened by U-boats. (See table 4.)

Table 4.—Relative Effectiveness of Alternative Aircraft Missions

Mission	Ships Saved/100 Sorties	
	M & K	S & T
Escort of Threatened Convoys	30	
Offensive Search in Bay of Biscay	5	10
Nonselective Convoy Escort	3	
Bombing of Biscay Ports	1	2

Neither Morse and Kimball nor Sternhell and Thorndike consider aircraft attrition. Surely the bombing of ports entailed greater attrition than the other missions, making it even less desirable than the above comparisons indicate.

Effectiveness of Search Radars Outside the Bay

The use of search radars and search receivers was not limited to the Bay of Biscay; land-based aircraft escorting convoys or coming to their rescue used radar, and surfaced U-boats at sea accordingly used their radar detectors. The operational sweep rate measure of search-radar effectiveness does not apply in this case, as the region searched is not well defined. A fairly direct measure of effectiveness is the exchange rate: U-boats sunk by convoy escorts per convoyed merchant vessel sunk. This quantity can be calculated from postwar records and is presented graphically in figure 19. Loss data for months with fewer than four sinkings of convoyed merchant vessels were not available.

This measure of effectiveness includes sinkings by escort vessels as well as by aircraft: the two kinds of platform worked together, and to evaluate the aircraft radar's effectiveness purely on the basis of sinkings by aircraft would constitute an oversimplification. The data[91] show a steady trend from a value of about 0.2 in January 1942 to about 0.8 in January 1944. Interestingly, the two high outliers, in July 1942 and May 1943, occur in months immediately following the deployment of new radars in the Bay of Biscay. We may surmise that new radars entered convoy escort service in these months, after the more needy Bay patrol aircraft had been equipped.

Figure 19
U-Boat/Merchant Exchange Rate

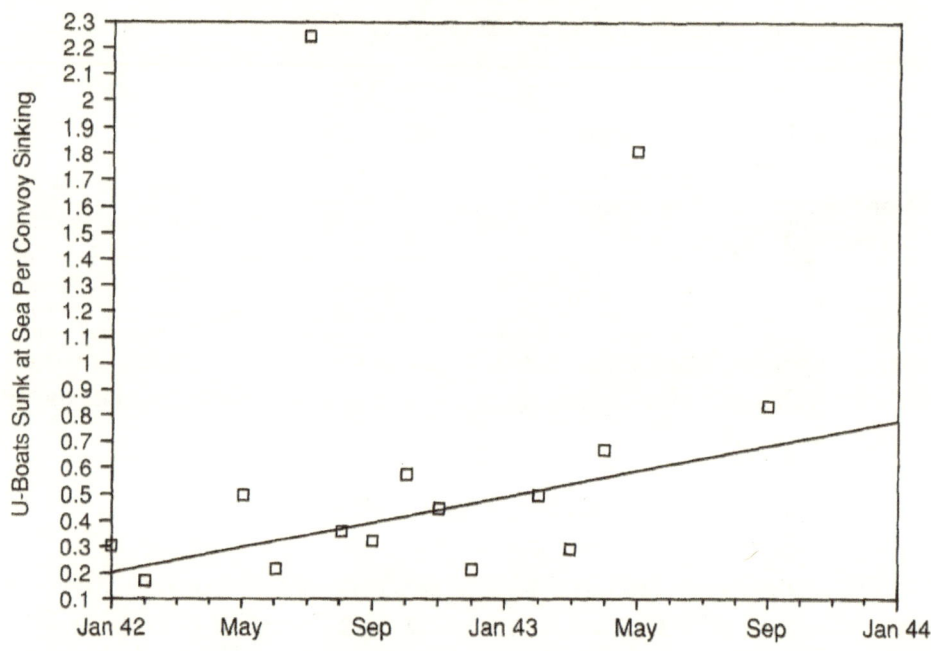

4

Modeling Search Operations in the Bay

The fact that . . . these expressions can be evaluated in closed form is a circumstance so remarkable as to cause one to suspect divine interference. The inverse cube law is therefore possibly holy, and in any case deserves consideration through being a compromise between the random and exhaustive assumptions, even in circumstances where the precise assumptions lying behind it are not directly verifiable.

—Alan R. Washburn, 1981[92]

So far we have quantified the interplay among the various tactics and pieces of equipment used by the two sides in the Bay, based only on results. The discussion has included a good deal of after-the-fact reasoning and little connection between cause and effect. It is a bookkeeper's analysis, not an economist's, comparing various static conditions rather than providing a dynamic model in which policy variables can be adjusted and the results observed. Further insight may come from modeling the engagement—working from ideas about the equipment to calculation of "predicted" sweep rates rather than working in the opposite direction, as we have up to now.

Such modeling will require some mathematical tools and submodels. In particular, we need to develop a way to handle the problem of assessing the chance that a searcher will detect a target before the target detects the searcher, given only the most rudimentary knowledge of the operational characteristics of the equipment used by the two sides. Operational sweep rate will again serve as a starting point.

In reflecting upon the opposing processes of detection and counterdetection, we quickly find that our sweep-rate conception of search does not quite capture the action we seek to model. The operational sweep rate worked well as a normalization of sightings according to flying effort, area searched, and submarines present. Now we want to work in the opposite direction, modeling the capabilities of pieces of equipment and then modeling their simultaneous action, so that we may ultimately ask the "what-if" questions generally barred from purely historical analysis.

Operational Sweep Width

As pointed out earlier, an airplane's operational sweep rate may be divided by its speed to get an operational sweep width. The notion of operational sweep width is not only formally correct (because square miles per hour divided by miles per hour equals miles) but also coherent: because the operational sweep rate is the rate at which ocean area would be examined by a perfect search aircraft as productive as an actual one, the operational sweep width is the width of the swath the perfect searcher would examine, given that it flies as fast as the actual aircraft.

An equivalent formulation defines the sweep width as the width of a swath such that the actual search overlooks as many targets (as always, we assume that these are randomly distributed) inside the swath as it sights beyond the swath. Because the swath extends to both sides of the airplane, one may equivalently say that sweep width is double the median lateral range to detected targets: the number of overlooked targets passing within half the sweep width of the airplane equals the number of detected targets lying beyond half the sweep width. In the parlance of atomic physics, the sweep width is "the capture cross-section of the airplane for U-boats."

Sweep Widths of Allied Detection Equipment

We now turn to the difficult task of finding a priori operational sweep widths for the equipment used in the Bay of Biscay. Most available accounts are of test-range performance, not operational use, and such test results are notoriously optimistic. In addition, visual ranges depended on weather conditions, and the range of any piece of equipment—including the human eye—depended strongly upon the altitude of the airplane from which it was used.

As mentioned earlier, the ASV Mark II L-band radar had a nominal range of up to 10 miles, and the ASV Mark III had a range of less than 15 miles. Morse and Kimball cite 10 miles as a "theoretical" range of the human eye as a detector of surfaced submarines, and estimate a factor-of-two real-world degradation attributable to "fatigue, etc."[93] A detailed table of Koopman's supports both numbers.[94] Koopman also points out that a spotter of a submarine or other small craft usually detects the wake rather than the vessel itself.[95]

The "realistic ranges" in tables 5 and 6 reflect the above estimates, including an across-the-board degradation by a factor of two, as cited for the human eye. Application of this factor to the radars is justifiable not only as standard treatment of test data—Waddington recommends a factor of two and a half[96]—but also by experience outside the Bay of Biscay, on the Moroccan Sea Frontier. Sternhell and Thorndike state as a rule that early radars did not outrange vision except at night or in bad weather, and that the ASG—the American S-band radar—had an operational sweep rate of 2,500 square miles per hour.[97] Assuming the standard airplane speed of 150 knots, the ASG then had a range of 8.3 miles: Waddington states that the ASG was 60 percent better than the ASV III, leading to a range estimate of 5.2 miles for the ASV III.

The ASV Mark III and the human eye also had to contend with snorkels as targets; Sternhell and Thorndike's estimates of these detection ranges are shown as well.[98] Because these estimates come from wartime experience and already reflect considerable degradation from test results given in the same source, no further factor of two was applied. They do not, however, come from sweep rates, so that we are in no danger of committing circular reasoning.

Table 5.—Realistic Detection Ranges

Detector	Realistic Ranges (miles)	
	vs. Submarine	vs. Snorkel
ASV Mark II	5	Not Used
ASV Mark III	5	0.05
Vision	5	0.3

Sweep Widths of German Warning Receivers

Just as a radar or visual lookout has a sweep width for a particular type of target, a warning receiver has a sweep width for a particular type of search radar: it will detect as many passing searchers beyond half the width as it will overlook within that distance. Unfortunately, quantifications of performance are even more difficult to find for radar warning receivers than for radars.

According to one source, the Metox "could detect an airborne [ASV Mark II] radar as much as ten to fifteen miles away."[99] Sternhell and Thorndike say that Vixen, the counter-countermeasure to Naxos, worked only if the contact initially appeared at a range of 15 miles or more, implying that Naxos had a range of at least 15 miles against the ASV Mark III.[100]* The estimate of a 2-mile visual-detection range for submarines' sighting of low-flying aircraft sweeps many human factors under the rug, but is a reasonable point estimate. Again, these numbers reflect operational, not test, experience, so no degradation factor applies. Nor are they based upon sweep rates, so we are not embarking upon a circular calculation.

Table 6.—Realistic Counter-Detection Ranges

Warning Device	Range (miles)
Metox	15
Naxos	15
Vision (day)	2

Lateral Range Curves

If we replace our mental image of a real aircraft, its bored crew scanning the gray ocean for U-boats, with one of a mathematically idealized aircraft

---

*   Naxos clearly evolved while in use. Morse and Kimball cite prisoner-of-war reports indicating a range of 8 to 10 miles, then say that development continued, mostly improving reliability through ruggedization, and say that "it eventually proved its value in giving warning of Allied S-band radar, usually at ranges about equal to contact ranges." They then provide the account of Vixen as given by Sternhell and Thorndike, indicating that Naxos by that time outranged the radar it received. (Morse and Kimball, p. 96.)

sweeping clean a swath of such a width that it contains as many submarines as the actual airplane sights, no harm results from supposing in the latter case that U-boats are always sighted when they are directly abeam of the airplane. Now let us recover some element of reality by recognizing that detection is a probabilistic, not a deterministic, phenomenon.

Suppose that the sighting of a submarine does not cause the aircraft to alter course, and that the crew's performance in their sighting is scored according to distance to the submarine when it comes abeam of the airplane instead of the distance at the moment of the sighting. We can then define the *lateral range curve*, which expresses the probability that the target will be sighted in terms of the target's distance from the airplane's ground track—that is, the separation between the target and the airplane when the target passes abeam of the airplane. (See figure 20.) In general, the probability depicted by the curve will be fairly high at the point of zero separation and then fall off to zero at some range beyond which sighting is impossible.

A perfect searcher such as that invoked in the description of operational sweep width would have a definite-range-law or "cookie cutter" lateral range curve, expressing a 100 percent probability of detection within half the lateral range and a zero probability of detection elsewhere. One does not normalize lateral range curves*: for any given searcher the area under the lateral range curve is the sweep width.

Knowing the detection capabilities of the Allied equipment would tell us the probability that a U-boat would be detected by a passing airplane—in a single glimpse or in a small slice of time—as a function of the range separating the two. We could then path-integrate the detection probabilities and arrive at probabilities of detection for various distances of closest approach. These probabilities, graphed, would form the lateral range curve of the aircraft; this curve would take into account any lessening of the airplane's effectiveness through counterdetection by the U-boat. Conversely, knowing the detection capabilities of the U-boat would tell us the probability that the U-boat sees the airplane first and submerges, *forestalling* detection by the aircraft.

In light of the differences of opinion as to the sweep widths of equipment used in the Bay of Biscay, one can hardly expect consensus regarding the exact probability of detection as a function of range.

---

*   Because they express probabilities, not probability densities; some searchers are just better than others.

Figure 20
Sweep Width Concept

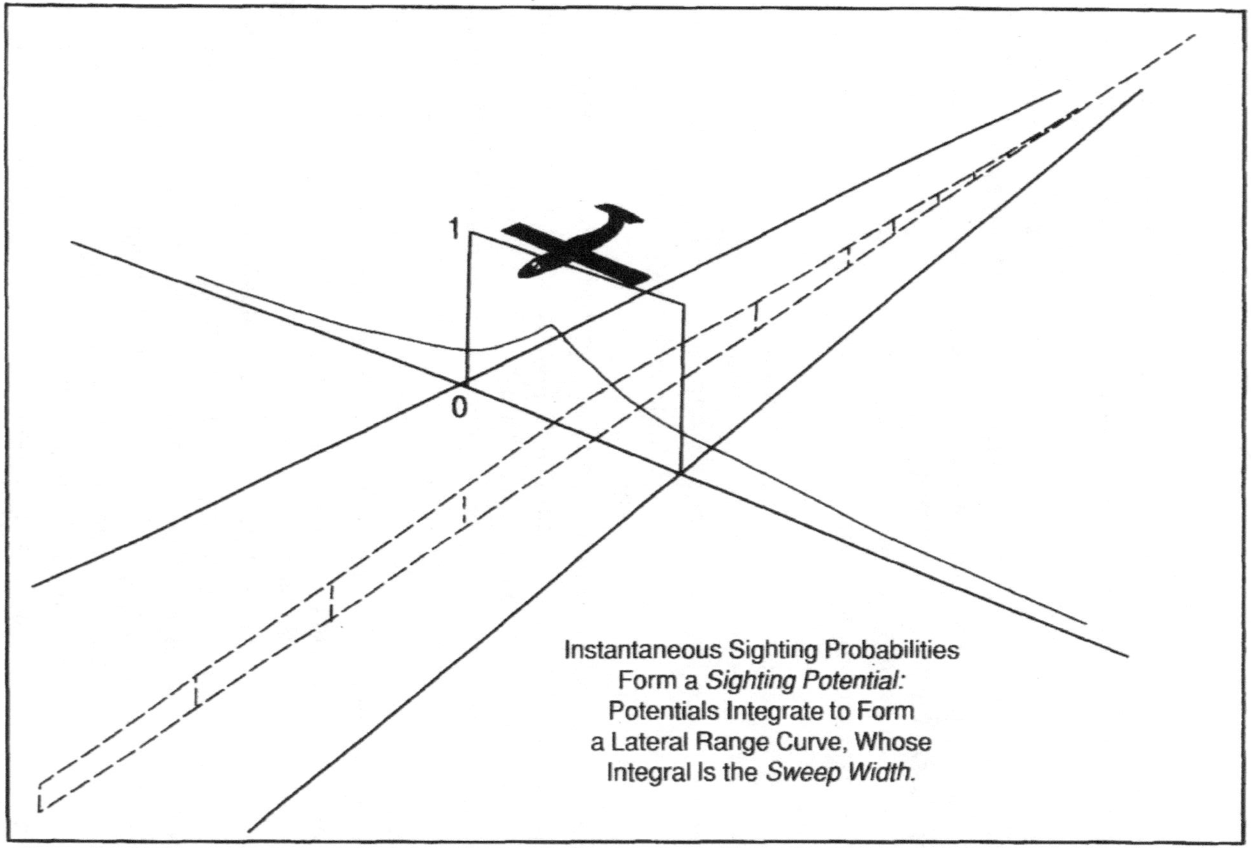

Instantaneous Sighting Probabilities
Form a *Sighting Potential*:
Potentials Integrate to Form
a Lateral Range Curve, Whose
Integral Is the *Sweep Width*.

A detection function that characterizes visual (and radar) detection, the so-called inverse cube law, is mathematically tractable enough that its lateral range curve can be found easily.[101] According to the inverse cube law, the probability of detection in any one glimpse directed at the target is inversely proportional to the cube of the range to the target.

In the case of vision, one might expect a single-glimpse probability density function inversely proportional to the square of the range, on the grounds that the solid angle subtended by the target so decreases. The extra factor arises because of the airborne observer's oblique perspective and the essentially horizontal nature of the target (the submarine's wake provides the strongest target for the eye[102]); distant targets are foreshortened more than closer ones seen from the same altitude. (See figure 21.)

In the case of radar, in which the submarine and not its wake is the dominant target, familiarity with the radar equation[103] could lead one to expect a fourth-power dependence of detection probability on range. However, nothing says that detection probability has to be proportional to returned signal energy (which does obey an inverse fourth-power law). Detection probability depends on the ratio of returned signal energy to the energy of the noise with which it is mixed, and on the ability of the radar operator to discern that a target is present. The operator's decision may depend on integration (by the radar display and by his own visual processes) over more than one radar scan; a pileup of closely spaced blips on the display therefore raises the detection probability. Pileups in azimuth will occur preferentially—all other things being equal—at longer ranges; an inverse-cube law may be derived by noting that this consideration inserts a range factor into the numerator of the radar equation.[104,105] Second World War experience commended using the inverse-cube law for radar.[106,107] (See figure 22.)

Koopman shows that when two detectors operate together, such as two observers looking out the same airplane window, their sweep width is not doubled, but multiplied by the square root of 2. In general, the simultaneous operation of n identical inverse-cube-type detection devices operated together multiplies the sweep width by the square root of n[108], so that

$$W_{total} = \sqrt{W_1^2 + W_2^2 + \ldots}$$

Figure 21
Inverse Cube Law for Vision

Viewed from above, distant, flat objects are **foreshortened,** leading to a third 1/r factor in subtended angle.

Figure 22
Inverse Cube Law for Radars

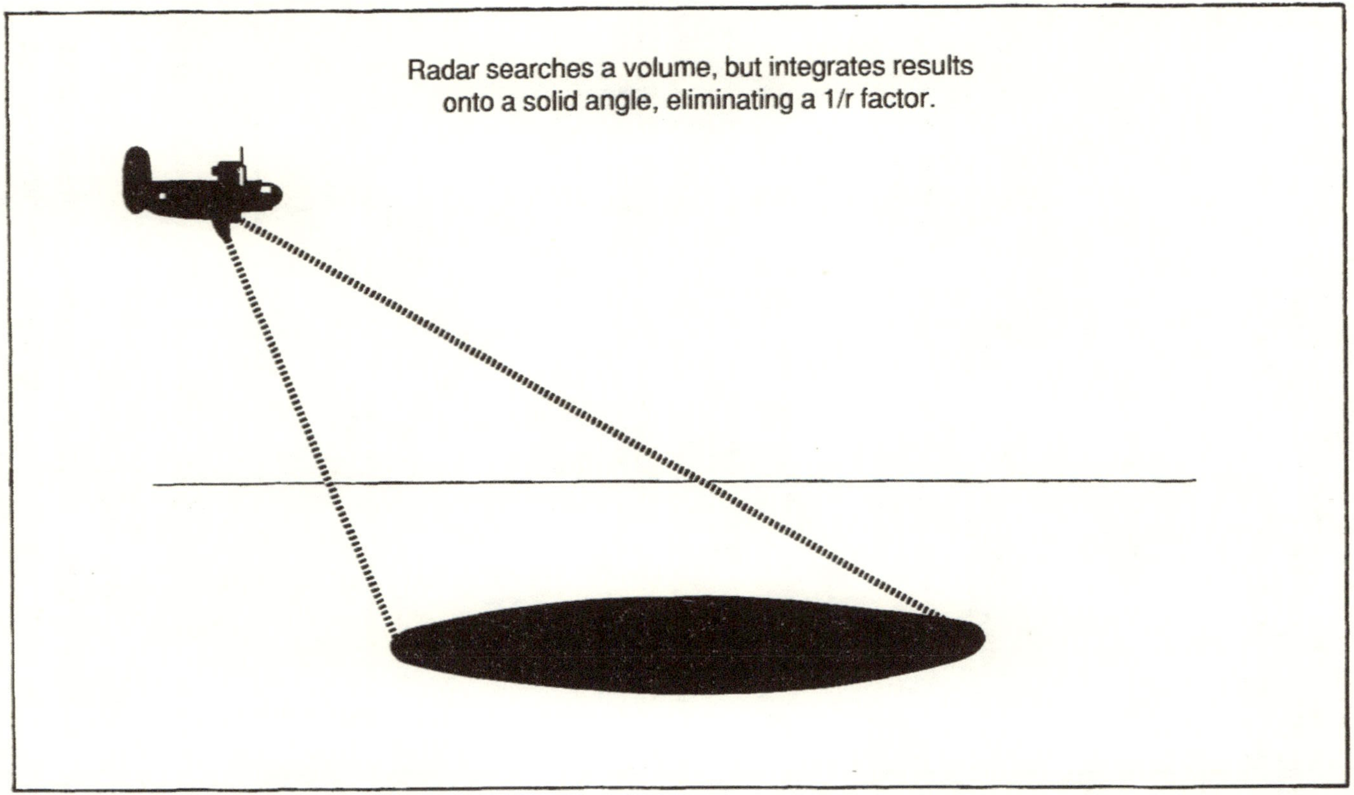

Radar searches a volume, but integrates results onto a solid angle, eliminating a 1/r factor.

MATHEMATICAL DETAILS

Following Koopman's notation, we may let $\gamma(\vec{r})$ be the instantaneous probability density of seeing a target at range $\vec{r}$. For a target with apparent motion past the searcher with speed v and distance of closest approach x, we may define the sighting potential F(x) as a path intregral over the (straight) flight path

$$F(x) = \oint \gamma(\vec{r}) \frac{ds}{v}$$

The probability that the target will be seen is then

$$p(x) = 1 - e^{-F(x)}$$

as one may prove by considering the probability of NOT seeing the target in any of a finite number of periods and then taking the limit as the number of periods, and their brevity, increases. The curve p(x) for all values of x is the lateral range curve. The sweep width W, then, is the width of a rectangle of unit height (i.e. p = 1) whose area is that under the lateral range curve.

$$W = \int_{-\infty}^{\infty} p(x)\, dx$$

In the case of the inverse cube sighting law,

$$\gamma(\vec{r}) = \frac{kh}{r^3}$$

where h is the altitude of the airplane and k is a constant containing all information about sighting conditions and the size of the target. The lateral range curve, found by first integrating to find F(x), is

$$p(x) = 1 - e^{-\left[\frac{W^2}{4\pi^2 \cdot x^2}\right]}$$

leading to a sweep width

$$W = 2 \cdot \sqrt{\frac{2\pi kh}{v}}$$

This result is, obviously, an approximation that breaks down when the number of detectors gets too large: no number of extra observers in an airplane would be able to spot U-boats over the horizon or beyond the meteorological limit of visibility. (Koopman characterizes this expression as the first answer to the question "How can the advantage of having n observers on the same searching aircraft instead of one be quantified?" as if to say that later, better, approximations were made.[109])

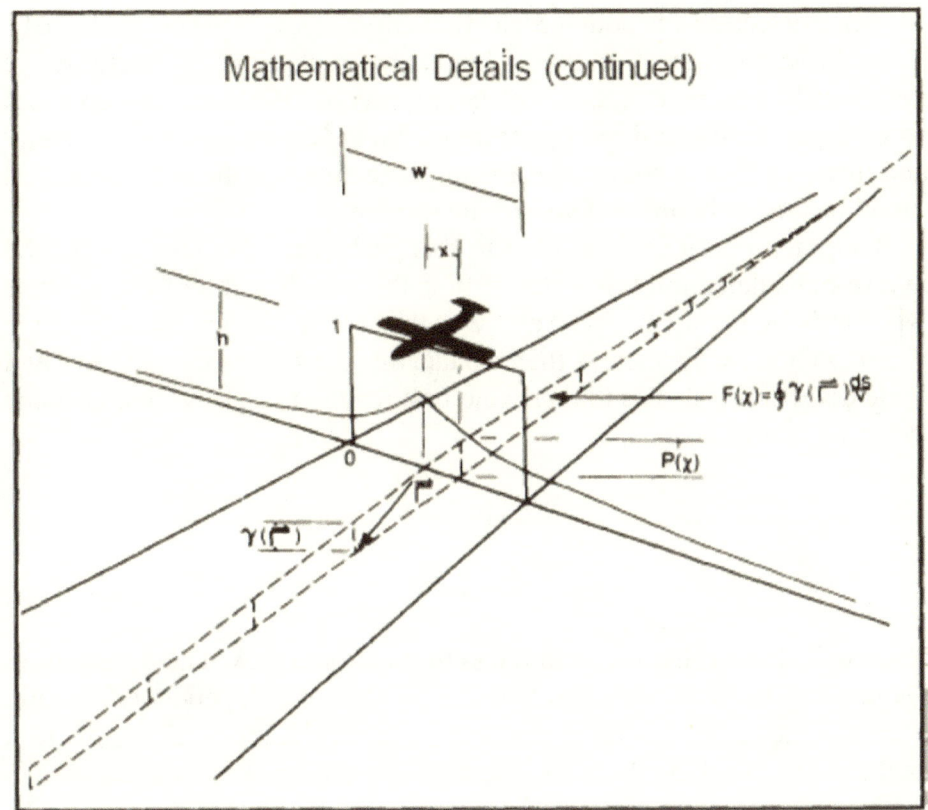

Mathematical Details (continued)

The effect of simultaneous visual and radar search can thus be quantified in terms of the resulting sweep width. Radar and visual sighting, each with a sweep width of 5 miles, combine to make an overall sweep width of roughly 7 miles (because 7 is approximately equal to 5 times the square root of 2).

These analyses of search operations apply equally well to a U-boat trying to detect an airplane.* The airplane, like a U-boat wake, presents a flat visual target lying in a horizontal plane; the search receiver, for reasons explained below, also operates according to an inverse-cube detection function. Assuming a typical search receiver with a sweep width of 15 miles and a typical lookout whose visual sweep width against aircraft is 4 miles, the submarine will have an overall sweep width of 15.5 miles.

---

*   In fact, they apply somewhat more intuitively, because the U-boat really wouldn't alter course to approach an airplane it had detected.

When detection and counterdetection efforts take place simultaneously, the cookie-cutter detection model begins to break down. If lateral range curves really were rectangular, whichever party had the longer sweep width would enjoy an unqualified upper hand. However, we know empirically (and intuitively) that sometimes the airplane detected the submarine and sometimes the submarine detected the airplane.

Koopman provides a solution to this problem.[110] He shows—for the inverse-cube detection law—that if W is the search width of the airplane (when looking for submarines) and W is the search width of the submarine (when looking for airplanes), then the fact that the submarine will dive and evade detection if it sees the airplane first results in an effective airplane sweep width of

$$\frac{W^2}{\sqrt{W^2 + W'^2}}$$

which will always be positive and less than or equal to W. This forestalling theorem can be generalized so that if a number of independent detection mechanisms contend with a number of independent counterdetection mechanisms, all of the inverse-cube type, the effective operational sweep width of the detecting platform is given by

$$\frac{W_1^2 + W_2^2}{\sqrt{W_1^2 + W_2^2 + \ldots + W_1'^2 + W_2'^2 + \ldots}}$$

where the independent counterdetection mechanisms might be vision and a radar detector, or perhaps the combined efforts of a group of submarines transiting together.

In order to create an overall formula for aircraft sweep width, taking into account visual and radar search and forestalling, we must establish that radar warning receivers, like visual search and radar, obey the inverse-cube law. Familiarity with the radar equation might lead one to believe that an inverse-square law would be more appropriate but, as noted above in discussing the radar case, there is no reason why the probability of detection has to be proportional to the received energy. In fact, one may more reasonably assume that the U-boat makes the detection when the received energy exceeds some threshold value. Because the aircraft is above water and shining its radar down obliquely, the locus of points on the water's surface at which a single radar scan will strongly illuminate the U-boat with energy is an ellipse. (See figure 23.) As the aircraft rotates its radar, this ellipse covers an annulus on the water's surface. Points near the outside of the annulus, however, are covered more briefly than points near the inside. Approximating the ellipse by a rectangle, we see that the exposure *time* is inversely proportional to the range, introducing an extra range factor in the denominator (where two range factors already exist, illumination *power* being inversely proportional to the square of the range) and leading to inverse-cube dependence of illumination *energy* on range.

Table 7 shows the operational sweep widths resulting from equipment parameters listed in tables 5 and 6 and the use of Koopman's forestalling theorem where applicable: Metox vs. ASV Mark II, Naxos vs. ASV Mark III, and vision vs. vision.

Submergence and the "Balanced Force" Concept

The most obvious submarine countermeasure of all is submergence. Submerged U-boats were virtually undetectable from the air, so if they spent half their time underwater they would in effect enforce a 50 percent reduction in sweep width. Sternhell and Thorndike provide a table showing the tradeoff between submergence and the time taken to transit the Bay. (See table 8.) Note that the faster transit requires disproportionately more surface time. The disproportion arises from time spent recharging the batteries used underwater—higher speeds consume not only more electricity per hour, but also more per mile.

Snorkeling would have permitted a passage of perhaps 75 hours with no time spent fully surfaced, but the snorkel did not appear until 1944.[111] The choice between daytime and nighttime search raised the question of how the Allies should apportion their search effort, especially after Metox made nighttime radar search so much less effective than daytime visual search. Ought patrol aircraft to fly by day, when the seeing is easy, or at night, when—because of the hostile daytime conditions—the U-boats are more likely to surface?

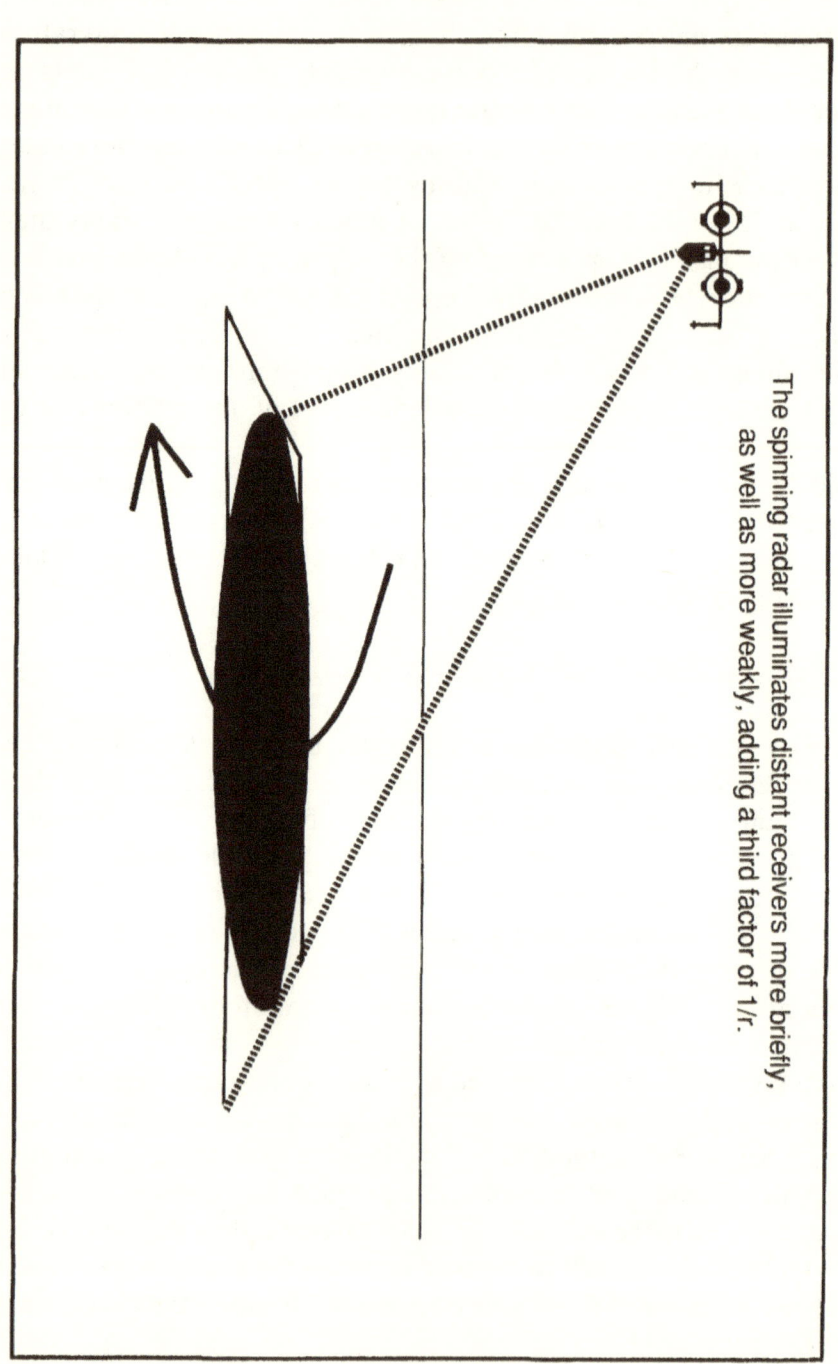

Figure 23
Inverse Cube Law of Receivers

The spinning radar illuminates distant receivers more briefly, as well as more weakly, adding a third factor of $1/r$.

**Table 7.—Sweep Widths as Degraded by Various Countermeasures
(in nautical miles)**

	Metox	Naxos	Darkness	Vision	Snorkel
ASV Mark II	3.2	10	10	9.3	Not Used
ASV Mark III	10	3.2	10	9.3	0.1
Vision	10	10	1	9.3	0.6
II & Vision	6.0	14.1	10	13.6	Not Used
III & Vision	14.1	6.0	10	13.6	0.6
	Vision and Metox		Vision and Naxos		
ASV Mark II	3.1		9.3		
ASV Mark III	9.3		3.1		
II & Vision	6.0		13.6		
III & Vision	13.6		6.0		

Table 8.—Tradeoff of Submergence and Bay Transit Time[112]

Hours on the Surface	Total Transit Hours
13	125
14	71
21	42
30	30

The situation recalls the story of the drunkard looking for his keys under a streetlamp. Asked where he dropped them, he points out into the surrounding gloom. Asked why he is looking under the lamp, he replies, "Because the light here is so much better."[113]

The U-boats, unlike keys, actively sought to evade detection. Under the "balanced force" concept propounded by Allied operations researchers,[114] flying effort was distributed so that an hour spent on the surface at night was as dangerous for the U-boat as an hour spent on the surface in the daytime. Any other partition of search effort would afford the U-boats the opportunity to surface only during the less dangerous period, thus reducing their overall danger. Conversely, the U-boats should have surfaced (and did[115]) at different times of the day and night, obliging the Allied aircraft to maintain around-the-clock coverage or, more precisely, making Allied concentration on any one part of the day fruitless. A policy of nighttime surfacing only, for example, would be an exploitable defect upon which the Allies could capitalize by abandoning daytime flight and searching only at night. Thus the situation is that of a saddle point in classic game theory, toward which the opponents gravitate. (See figure 24.)

The balanced-search approach ignores the possibility of danger to aircraft; if such danger existed and were different by day and night, considering airplane attrition would lead planners to depart somewhat from the saddle-point solution derived above. In fact, enemy danger to aircraft was negligible by day and zero by night.[116] U-boats fought back with FLAK for only a brief period, and the Luftwaffe seldom could spare enough aircraft to provide significant aid to the Kriegsmarine in the Bay of Biscay; indeed, German attempts to provide fighter cover for U-boats were for a time counterproductive, because they cued the Allies as to the days when U-boat traffic would be heavy.[117] In fact, out of 9,000 sorties flown, the Bay patrol lost only about 50 aircraft to enemy action, and was never seriously hampered by the threat of enemy air action.[118] Many more aircraft were lost to accidents;[119] one may safely assume that accidents were more common at night, compensating for the air threat by day *if* aircraft attrition was a significant consideration.

The search-width concept allows us to find the correct partition of flying effort between day and night.[120] Let W_d and W_n be the daytime and nighttime sweep widths of an aircraft, F_d and F_n be the daytime and nighttime amounts of flying (in hours), and D and N be the number of hours of daylight and night in a given 24-hour period. Neither W_d nor W_n is zero, or else there would be no allocation problem. F_d and F_n add up to F, the total amount of flying effort available; D and N, of course, add up to 24. If an hour spent on the surface is to be as dangerous at night as by day, then flying effort must be so arranged that

$$\frac{F_d \cdot W_d}{D} = \frac{F_n \cdot W_n}{N}$$

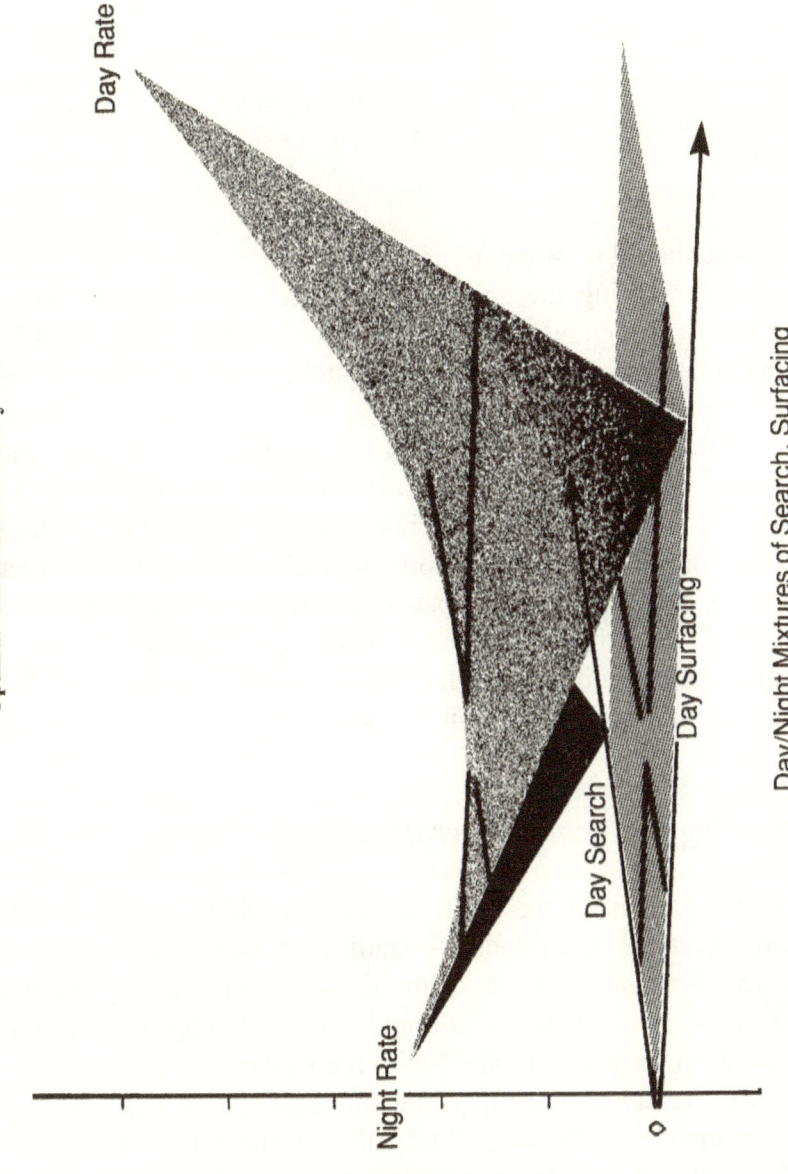

Figure 24
Saddle Point in "Balanced Search"
"Optimal" Mix Nullifies Enemy Mix

Day Rate

Day Surfacing

Day Search

Night Rate

Day/Night Mixtures of Search, Surfacing

Sweep Rate, Square Miles/Hour

provided that the region under surveillance is so large as to preclude any effect of diminishing returns. We can then solve for the daytime fraction of flying effort:

$$\frac{F_d}{F} = \cfrac{1}{1 + \cfrac{W_d \bullet (24-D)}{W_n \bullet D}}$$

The rest should be spent at night.

Implementing the above solution requires accurate knowledge of the true daytime and nighttime sweep widths: any error would create an imbalance between danger at night and by day, which the U-boats could then exploit—if they knew about it—by surfacing only during the less hazardous period. This solution also requires a fungible force whose daytime flying hours can convert on a one-for-one basis to nighttime flying. In fact, Sternhell and Thorndike cite nighttime surfacing as safer in the summer of 1942 (before the enunciation of the balanced-force concept) because only a few patrol aircraft carried equipment for nighttime search.[121] For their part, the Germans alternated between preferring daytime and nighttime surfacing, showing that they felt the intended "damned if you do and damned if you don't" effect of the balanced-force approach.

Modeling Bay Search Operations

By using what we know about both sides' equipment and tactics (particularly U-boats' tactics—trading speed for submergence) we may use the foregoing tables and equations to estimate the operational sweep rate aircraft "should have" attained in the Bay. Figure 25 shows the actual Bay transit time as compared with the model.

We already have the daytime and nighttime sweep widths W_d and W_n determined by each side's choice of equipment and the partition of flying effort F into daytime and nighttime portions F_d and F_n determined in turn from the widths and the principle of balanced search. These widths must be degraded to reflect the tactics of the U-boats; submerged U-boats were practically impossible to sight from airplanes.

Figure 25
Bay Transit Time: Actual vs. Model Input

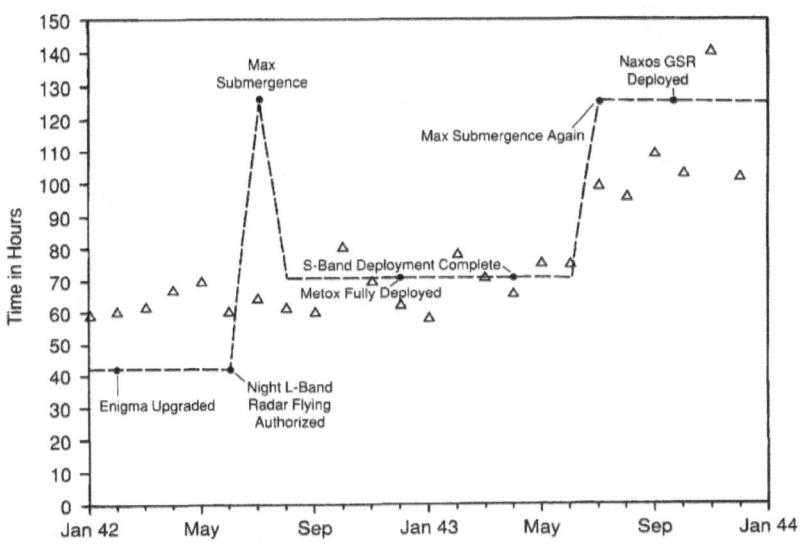

● *Line Equals Model Input Values; Symbols Equal Actual Values.*

We will keep track of whether the U-boats prefer to surface by day or by night; even though the balanced search neutralized this choice, the Germans considered the preferred surfacing time a policy variable. In any case, the distinction facilitates computation of sweep rates. Given the policy choice, the total transit time T, the total time spent on the surface S (as in table 9), and the durations of day and night D and N, we can write down the number of hours spent on the surface.

With a policy of surfacing only at night if possible, the U-boat spends either S hours or $T \cdot N/(N+D)$ hours—whichever is less—on the surface by night. The latter quantity is the number of nighttime hours during a transit of T hours started at a random time of day. (The assumption of randomness is better for incoming than for outgoing U-boats, which probably tried to coordinate their departures with the sun as well as with the tides.) If

$$S > \frac{T \cdot N}{N + D}$$

the extra time $S - T \cdot N/(N+D)$ will have to be spent on the surface by day.

Table 9.—Data Regarding Daylight and Surfacing in the Bay (daylight,[122] surfacing[123])

Month	Hours of Daylight	Hours of Night	Time on Surface (Hours)	Month in Hours	Period on Surface	Transit Time (Hours)
Jan 42	8.50	15.50	21	744	night	42
Feb	10.12	13.88	21	672	night	42
Mar	11.80	12.20	21	744	night	42
Apr	13.73	10.27	21	720	night	42
May	15.37	8.63	21	744	night	42
Jun	16.35	7.65	21	720	night	42
Jul	15.63	8.37	13	744	day	125
Aug	14.55	9.45	14	744	night	71
Sep	12.70	11.30	14	720	night	71
Oct	10.78	13.22	14	744	night	71
Nov	9.10	14.90	14	720	night	71
Dec	8.08	15.92	14	744	night	71
Jan 43	8.50	15.50	14	744	night	71
Feb	10.12	13.88	14	672	night	71
Mar	11.80	12.20	14	744	night	71
Apr	13.73	10.27	14	720	day	71
May	15.37	8.63	14	744	day	71
Jun	16.35	7.65	14	720	day	71
Jul	15.63	8.37	13	744	day	125
Aug	14.55	9.45	13	744	night	125
Sep	12.70	11.30	13	720	night	125
Oct	10.78	13.22	13	744	night	125
Nov	9.10	14.90	13	720	night	125
Dec	8.08	15.92	13	744	night	125
Jan 44	8.50	15.50	13	744	night	125

Similarly, with a policy of surfacing only by day if possible (as under the fight-back-with-FLAK policy), the U-boat spends either S hours or $T \cdot D/(N + D)$ hours, whichever is less, on the surface by day, with any extra time

$$S - T \cdot D/(N + D), \text{ if } S - T \cdot D/(N + D) > 0$$

spent on the surface at night.

With the surfacing time S thus partitioned into daytime and nighttime portions S_d and S_n (one of which may be zero—both if the snorkel is used), the daytime and nighttime sweep widths W_d and W_n are multiplied by S_d/T and S_n/T, respectively, because submarines can be sighted only during the surfaced portions of their transits. Table 10 shows a month-by-month day-night breakdown of the exposure of U-boats to Allied detection, the Allied detectors in use, and the devices used by the Germans to forestall detection.

Table 10.—Exposure Times in Hours per Transit, Allied and German Equipment

Roman Numerals Denote Marks of ASV Radar

Month	Exposure		Allied Device		German Device	
	day	night	day	night	day	night
Jan 42	0.0	21.0	vision	I	vision	darkness
Feb	0.0	21.0	vision	I	vision	darkness
Mar	0.0	21.0	vision	I	vision	darkness
Apr	3.0	18.0	vision	I	vision	darkness
May	5.9	15.1	vision	I	vision	darkness
Jun	7.6	13.4	II & vision	II	vision	darkness
Jul	13.0	0.0	II & vision	II	vision	darkness
Aug	0.0	14.0	II & vision	II	vision	darkness
Sep	0.0	14.0	II & vision	II	vision	darkness
Oct	0.0	14.0	II & vision	II	vision/Metox	Metox
Nov	0.0	14.0	II & vision	II	vision/Metox	Metox
Dec	0.0	14.0	II & vision	II	vision/Metox	Metox
Jan 43	0.0	14.0	II & vision	II	vision/Metox	Metox
Feb	0.0	14.0	II & vision	II	vision/Metox	Metox
Mar	0.0	14.0	II & vision	II	vision/Metox	Metox
Apr	14.0	0.0	III & vision	III	vision	darkness
May	14.0	0.0	III & vision	III	vision	darkness
Jun	14.0	0.0	III & vision	III	vision	darkness
Jul	13.0	0.0	III & vision	III	vision	darkness
Aug	0.0	13.0	III & vision	III	vision	darkness
Sep	0.0	13.0	III & vision	III	vision	darkness
Oct	0.0	13.0	III & vision	III	vision/Naxos	Naxos
Nov	0.0	13.0	III & vision	III	vision/Naxos	Naxos
Dec	0.0	13.0	III & vision	III	vision/Naxos	Naxos
Jan 44	0.0	13.0	III & vision	III	vision/Naxos	Naxos

Multiplying the sweep width (in miles) by the speed of the airplane (in miles per hour) gives a figure for the sweep rate (in square miles per hour) of the airplane:

sweep width (miles) • speed (mph) = sweep rate (sq. miles/hour).

Recalling that the total flying effort F is partitioned into daytime and nighttime portions F_d and F_n, we can express the total operational sweep rate as

$$V \bullet \left\{ \frac{F_d}{F} \bullet W_d \bullet \frac{S_d}{T} + \frac{F_n}{F} \bullet W_n \bullet \frac{S_n}{T} \right\}$$

where V is the airplane's speed and the quantity in braces is the sum of the daytime and nighttime search widths, weighted according to the partition of flying effort. Again, the policy of balanced search will force the two terms inside the braces to be equal. See table 11.

Table 11.—Sweep Widths (mi.), Effort, and Resulting Raw Sweep Rate (mi.^{2}/hr.)

Month	Sweep Widths		Sweep Effort		Resulting
	day	night	day	night	Rate
Jan 42	9.3	2.0	11%	89%	134
Feb	9.3	2.0	14%	86%	130
Mar	9.3	2.0	17%	83%	124
Apr	9.3	2.0	22%	78%	122
May	9.3	2.0	28%	72%	132
Jun	13.6	10.0	61%	39%	412
Jul	13.6	10.0	58%	42%	123
Aug	13.6	10.0	53%	47%	139
Sep	13.6	10.0	45%	55%	162
Oct	6.0	3.2	30%	70%	65
Nov	6.0	3.2	24%	76%	71
Dec	6.0	3.2	21%	79%	74
Jan 43	6.0	3.2	22%	78%	73
Feb	6.0	3.2	28%	72%	68
Mar	6.0	3.2	34%	66%	62
Apr	13.6	10.0	50%	50%	200
May	13.6	10.0	57%	43%	228
Jun	13.6	10.0	61%	39%	246
Jul	13.6	10.0	58%	42%	123
Aug	13.6	10.0	53%	47%	73
Sep	13.6	10.0	45%	55%	85
Oct	6.0	3.2	30%	70%	34
Nov	6.0	3.2	24%	76%	37
Dec	6.0	3.2	21%	79%	39
Jan 44	6.0	3.2	22%	78%	38

Saturation and the "Law of Random Search"

To go from the search effectiveness of an airplane to that of the entire Bay patrol fleet requires an estimate of how efficiently the airplanes' search efforts were coordinated. We can estimate fleet effectiveness for two cases: total coordination and total incoordination. Actual effectiveness should fall between these bounds.

Multiplying the sweep rate by the number of patrol hours flown by the whole force in a month leads to an upper bound on the effectiveness of the whole force. Finding the area swept

$$\text{sweep rate (mi}^2\text{/hr)} \bullet \text{flying effort (hrs)} = \text{area swept (sq. mi.)}$$

leads to the number of submarines destroyed, given the probability P that a sighted U-boat will be sunk:

$$\text{area swept} \bullet \text{sub density} \bullet P = \text{sinkings.}$$

Equivalently, we may consider the ratio of the swept area to the area of the whole Bay to be the probability that the submarine is sighted, leading to

$$\text{subs} \bullet \text{(area swept/Bay area)} \bullet P = \text{sinkings.}$$

This figure, as mentioned above, is an upper bound—that is, an overestimate in some cases and an underestimate in none. (The model is illustrated in the left half of figure 26.) The overestimate comes from having treated the sweep width as being in fact a width swept clean of submarines, when that treatment is really just an accounting convention introduced to cope with the probabilistic nature of detection. (Recall that the sweep width is the width of the swath an airplane would sweep clean if it detected in a clean sweep the same number of submarines as it actually detects.) Because the airplanes really perform an imperfect search over a larger region, their efforts overlap to some degree, lessening the number of submarines sighted or killed.

Figure 26
Confetti Search Concept

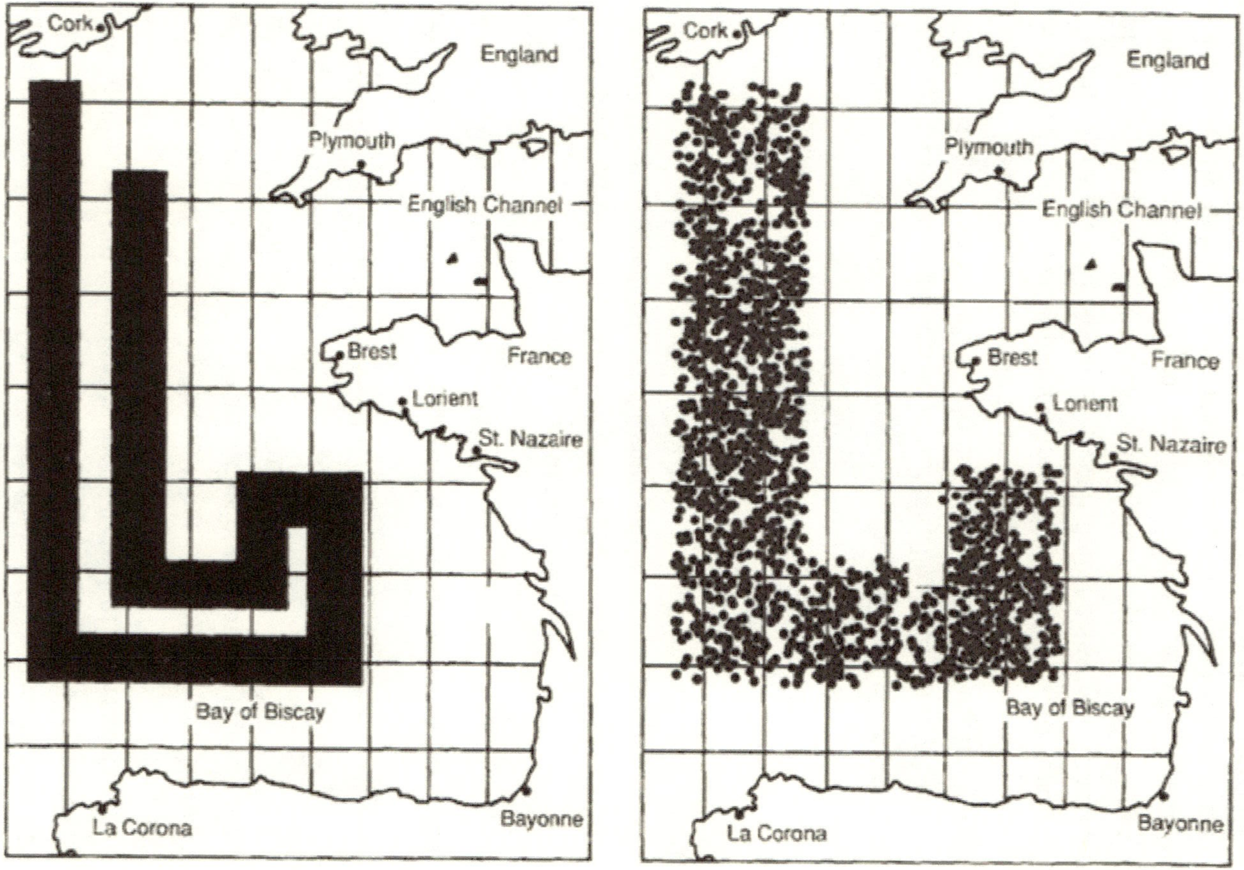

This difficulty becomes most apparent in the case in which the area swept exceeds the total area of the Bay.

We may also construct a lower bound for the expected number of sinkings—an underestimate in some cases and an overestimate in none. As above, multiplying the swept area by P leads to what might be considered the lethal area of the search—the number of square miles' worth of submarines that it kills off. The upper bound calculation tacitly assumed that this lethal area could be neatly arranged in the Bay, with no overlaps wasting lethality by sinking the same submarine twice. The opposite extreme would resemble a haphazard crop-dusting operation that randomly splatters lethality, as in the right half of figure 26. Washburn describes such a search, called a "random search" by Koopman,[124] as a "confetti search":

> In effect, the searcher cuts [the lethal area] into confetti and drops it at random—The complete lack of organization inherent in attempting to search by confetti casting explains why random search is often treated as a lower bound in computing detection probability, even though a deliberate attempt to create a low (!) detection probability (for example by not moving) could do even worse.[125]

We may find the area actually rid of submarines (corresponding to the area covered by the confetti) by viewing the lethal area A (equal to $P \cdot$ area swept) as cut up into n equal pieces of confetti. Given a Bay of area B onto which the confetti is dropped, a particular point remains uncovered with probability $(1 - A/nB)^n$. For large values of n, this chance approaches $e^{-A/B}$, with a complementary $1 - e^{-A/B}$ chance that the confetti covers the point.* This approach leads to the following formula for submarines sunk:

$$\text{subs sunk} = \text{subs present} \cdot (1 - e^{-P \cdot \text{area swept/Bay area}})$$

which contrasts with yet a third formulation for the upper bound:

$$\text{subs sunk} = \text{subs present} \cdot (P \cdot \text{area swept/Bay area}).$$

---

* This formula can be thought of as the chance of not getting zero hits in a Poisson distribution of hits h where $P(h) = (A/B)^h e^{-(A/B)}/h!$.

Figure 27
Orderly vs. Random Search

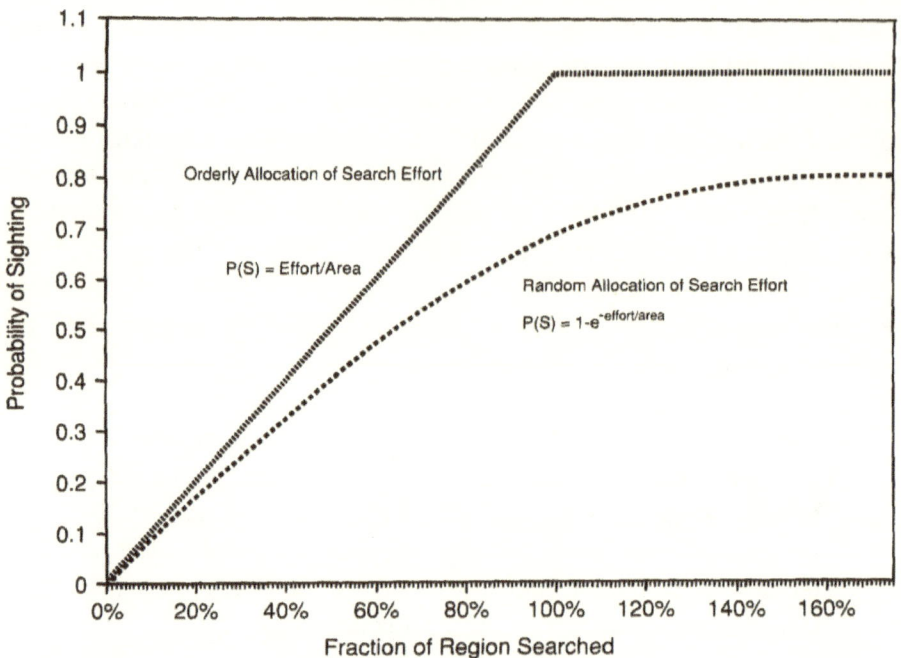

One may usefully consider the lower bound as the expected value under a Poisson distribution with density equal to

$$P \bullet \text{area swept/Bay area}$$

and the upper bound as the expected value under a Bernoulli distribution with the same density.

At low densities, the Bernoulli distribution approaches the Poisson,[126] as one can see by considering the series expansion. Thus the lower bound on submarines sunk tends toward the upper bound for values of the area swept which are small compared to that of the whole Bay. (See figure 27.) This convergence agrees with intuition: little overlap results when a small quantity of confetti is cast onto a large floor.

Another convergence, which will be useful in the following section, arises in comparing historical sweep rates with the rate from the model. Because the historical sweep rate is calculated on the basis of results, using

$$\text{sweep rate} = (\text{sightings/U-boat density})/\text{flying hours}$$

it already contains the effects of any saturation. The rate does not refer to P, the probability of kill. We may obtain a comparable sweep rate from the model by taking

$$\text{sweep rate} = \frac{1 - e^{-P \cdot \text{area swept/Bay } area}}{P \cdot \text{flying hours/Bay area}}$$

Table 12.—Effect of Lethal-Area Saturation on Bay Sweep Rates

Month	Sweep Rate	Flying Hours	Actual P(K|S)	Moving Av P(K|S)	Rate with Saturation
Jan 42	134	350	0%	0%	134
Feb	130	500	0%	0%	130
Mar	124	400	0%	0%	124
Apr	122	800	0%	0%	122
May	132	1000	0%	2%	131
Jun	412	2600	0%	3%	372
Jul	123	3750	10%	3%	116
Aug	139	3200	3%	4%	129
Sep	162	4100	3%	4%	146
Oct	65	4100	6%	2%	64
Nov	71	4600	0%	2%	69
Dec	74	3400	0%	2%	73
Jan 43	73	3130	0%	1%	72
Feb	68	4400	3%	1%	66
Mar	62	4600	2%	3%	60
Apr	200	4200	2%	4%	175
May	228	5350	7%	7%	168
Jun	246	5900	7%	21%	96
Jul	123	8700	16%	22%	57
Aug	73	7000	71%	21%	50
Sep	85	8000	10%	25%	47
Oct	34	6000	0%	23%	29
Nov	37	7000	29%	12%	33
Dec	39	6000	7%	12%	35
Jan 44	38	5000	14%	16%	34

The numerator is the fraction of the Bay covered at least once by the lethal effects of search. Division by P yields an effective rate of detecting (not destroying) per hour of flight. Division by the number of hours flown yields the fraction of the Bay searched at least once in an hour of effort, and then multiplication (that is, division of the denominator) by the area of the Bay finally gives a sweep rate in square miles per hour. Difficulty will arise, however, because in some months the historical P is zero: no U-boats were sunk in the Bay despite some sightings. If the reader's intuition about the mathematics of search operations has been sufficiently exercised up to this point, he or she may suspect that the lower bound of the sweep rate converges toward the upper bound for decreasing values of P even as it did for decreasing values of the area swept. This intuition is correct, though less easily proved than the former result because of the presence of P in the denominator. Application of l'Hôpital's Rule clears that obstacle and shows that, indeed, the no-overlap formula applies in cases of small or zero P.

Comparison of the Model With Reality

As figure 28 shows, the constructed sweep rates agree well with the actual ones presented earlier. The model overstates the early effect of Metox because it assumes that all the U-boats suddenly got Metox all at once, whereas in fact the deployment took several months. The model's estimate of Metox effectiveness agrees extremely well with the sweep rates of winter 1942-1943, with Metox fully deployed and the S-Band ASV III not yet introduced. Sternhell and Thorndike examine contemporaneous Metox performance against ASV Mark II in the Trinidad area and find no such improvement.[127] They conclude that either seasonal or psychological factors were at work in the Bay. However, they were working with wartime data that understated the number of submarines in the Bay. Paradoxically, such an underestimate of the submarines' density total leads—regardless of the number of transits—to an overestimate of the density of surfaced submarines: a lesser density total implies that each submarine spends less time in transit, presumably because it spends more time surfaced, and thus moves faster.

The model's agreement with reality would improve if we knew a general relation between the amount of time spent on the surface during a transit of the Bay and the amount of time required for the whole transit. Lacking

Figure 28
Actual vs. Calculated Sweep Rates

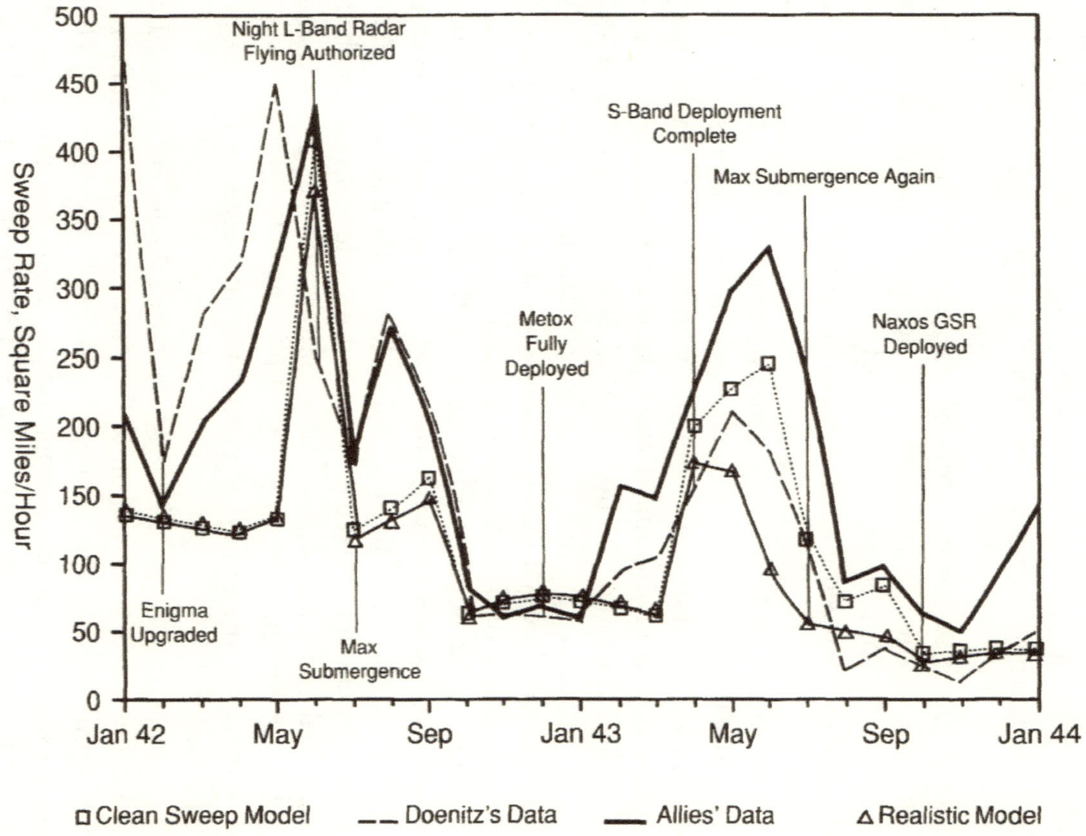

such a relation, we must use the handful of values given by Sternhell and Thorndike, determining time spent on the surface as closely as possible from the transit times computed from the Dönitz *War Diary*. (The average transit time for a given month comes from dividing the number of U-boat days spent in the Bay by the number of transits.)

The only seasonal factor the calculation embraces is that of variation in the daily amount of daylight. Because nighttime sweep widths are smaller than daytime ones, winter's longer nights make surfacing safer—at all hours, if the balanced-force concept is in play—allowing for a faster transit. Other seasonal factors not accounted for in the model, such as changing daytime ranges for visual detection of submarines, may also be at work. We lack the data needed for an accurate assessment of the effect on sweep width of seasonal factors other than the changing length of the day.

Sternhell and Thorndike mention, as a possible psychological benefit of Metox, the renewed (and correct, because of the lack of night-flying aircraft—evidently the balanced-force idea was not yet fully implemented) feeling that it was safe to surface at night. They suppose that the initial introduction of ASV Mark II caused a switch to the more dangerous practice of daytime surfacing justified, wrongly, on the grounds that in daylight the submarine could at least counterdetect the aircraft and dive. This practice, ordered by Dönitz on June 17, 1942,[128] resulted in the large positive spike (both calculated and observed) in operational sweep rate when the ASV Mark II was first introduced. The later reversion to nighttime surfacing lowers the calculated effectiveness, but not by enough to get it down to historical levels, possibly because the historical operations did not fully comply with the yet-to-be-enunciated balanced-search principle. The July 1942 dip comes from another overreaction, the policy of maximum submergence Dönitz mandated on June 23.[129]

Part of the sweep-width model can be validated in a second way. The reduction due to Naxos of the ASV Mark III sweep width from 10 miles to 3.2 miles corresponds extremely well with Sternhell and Thorndike's observation—based on disappearing contacts and previous ASV Mark III performance in the Bay—that with Naxos in use patrols sighted just 31 percent (46 out of 149 sightings expected, on the basis of previous experience, during the period of Naxos use) of the U-boats that would have been sighted had Naxos not been used.[130] The otherwise-derived figures for the sweep widths of the ASV Mark III and the Naxos lead to the same result when Koopman's forestalling theorem is applied. Before Naxos, the ASV Mark III had a sweep width of 10 miles; Koopman's forestalling theorem and the 15-mile range imputed to Naxos result, as we have seen, in a 3.2-mile width for the ASV Mark III when operated against Naxos. Thus the aircraft would spot 32 percent of the U-boats that would have been sighted had Naxos not been used, a result in remarkable accord with the 31 percent estimated by Sternhell and Thorndike. The agreement provides empirical verification of the use of the inverse-cube law for search receivers.

Sternhell and Thorndike partition the remaining 69 percent into 65 cases (44 percent) in which the ASV Mark III acquired a target but lost it before visual contact could be made, and 38 cases (25 percent) in which Naxos detected the ASV Mark III and enabled the U-boat to dive before being detected at all.[131] The first category, "disappearing contacts," is a matter of record, but the frequency of the second type of event could only be inferred from the aggregate decrease in contacts after the introduction of Naxos.

Alternatively, we may consider Naxos to be an airplane detector whose performance is sometimes forestalled by the airplane's radar. The degraded sweep width of Naxos is then

$$\frac{W_N^2}{\sqrt{W_N^2 + W_A^2}}$$

by Koopman's forestalling theorem. This is the Naxos's sweep width for airplanes that have not detected the Naxos-carrying U-boat. Contacts that the airplane reports as "disappearing" result from cases in which Naxos detects an airplane that has seen the U-boat. The unit's *disappearing-contact sweep width* is thus

$$W_N - \frac{W_N^2}{\sqrt{W_N^2 + W_A^2}}$$

As we shall see in the following section, this formulation squares well with experience.

The False-Alarm Sweep Width

Consideration of the lateral-range curves $P_{ASV}(r)$ and $P_{Naxos}(r)$* shows that for any range r there is a probability

$$(1 - P_{ASV}(r)) \bullet P_{Naxos}(r)$$

that a U-boat will detect an airplane passing at range r without the airplane detecting the U-boat. In such a case, the U-boat cannot be said to have forestalled a sighting by the aircraft: sighting—a probabilistic event—just happens not to occur even if the U-boat stays on the surface. From the Naxos operator's point of view, the event is a "false alarm." Like all false alarms, its true nature can be known only in retrospect, if then; a U-boat skipper who elects to stay on the surface and goes unattacked would know that a false alarm had occurred, but one who submerges would never know whether the alarm had been false.

False alarms posed a problem for the U-boats because a sudden dive not only interrupted the recharging process but used up a significant amount of battery energy: to replace the energy used in the dive alone required 30 minutes of running on the surface by diesel.[132] (After the dive, boats were ordered to stay underwater for about half an hour,[133] using up additional battery energy.) During the period in which Metox search receivers countered ASV Mark II radars, Allied operations researchers urged the deliberate creation of false alarms through "flooding," in which specially equipped aircraft spread radar-like signals over a wide region, causing U-boats to spend extra time on the surface recovering from crash dives. The very brief trial of this technique showed that it had some merit.[134]

Given our not-unfounded assumption that the ASV Mark III and the Naxos obeyed the inverse-cube detection law, we can derive a *false-alarm sweep width* for the Naxos, and then an expected value for the number of false alarms per Bay transit.

---

*   As the reader will recall, the lateral range curve characterizes a detector (a radar or a search receiver) in terms of its probability of detecting a target that is a distance r away as of its point of closest approach.

False Alarm analysis of ASV III vs. Naxos in the Bay

H = 1831 — total nighttime hours in the period of interest (Nights in October, November, December, and January)
F = 18160 — total flying hours in the period of interest
B = 130000 — area of the Bay
T = 13 — surfaced hours/transit of a U-boat
S = 150 — aircraft speed
Wa = 10 — undegraded sweep width of ASV III
Wn = 30 — sweep width of Naxos

ITEM	FORMULA AND THEORETICAL VALUE	HISTORICAL VALUE
Sightings with Naxos	$\dfrac{T\,F\,S}{H\,B} \cdot \dfrac{Wa^2}{\sqrt{Wa^2 + Wn^2}} = 0.47$	$\dfrac{46}{281} = 0.164$
Degradation due to Naxos	$\dfrac{Wa}{\sqrt{Wa^2 + Wn^2}} = 0.316$	$\dfrac{46}{149} = 0.309$
Naxos alarms	$\dfrac{T\,F\,S}{H\,B} \cdot Wn = 4.463$	not available
Naxos first-detections i.e. forestallings & false alarms	$\dfrac{T\,F\,S}{H\,B} \cdot \dfrac{Wn^2}{\sqrt{Wa^2 + Wn^2}} = 4.234$	not available
Naxos false alarms	$\dfrac{T\,F\,S}{H\,B} \cdot \left[\sqrt{Wa^2 + Wn^2} - Wa \right] = 3.217$	not available
Disappearing Contacts, i.e alarms minus 1st detections	$\dfrac{T\,F\,S}{H\,B} \cdot \left[Wn - \dfrac{Wn^2}{\sqrt{Wa^2 + Wn^2}} \right] = 0.229$	$\dfrac{65}{281} = 0.231$

This interaction of radar and search receiver is easier to attack mathematically than was forestalling because the lateral-range-curve idea can be applied (see box), and the resulting formula is simple:

$$W_{fa} = \sqrt{W_{Naxos}^2 + W_{ASV}^2} - W_{ASV}$$

where W_{fa}, W_{Naxos}, and W_{ASV} are the false-alarm, Naxos, and (unforestalled) ASV Mark III sweep widths, respectively. Given the values used earlier for the ASV Mark III and Naxos sweep widths (10 and 30 miles, respectively), the false-alarm sweep width of the Naxos receiver turns out to be 22 miles.

With these sweep widths in hand, we can readily calculate the expected number of ASV contacts and false alarms experienced by a U-boat making the Bay passage. During the period of Naxos use, U-boats surfaced only at night, so the definitions in table 13 provide the correct basis for the calculations.

Table 13.—Basis for Calculating Occurrence of Naxos False Alarms

Basic Variable	Definition	Value, 10/43 – 1/44
H	Hours of Darkness	1,830
F	Nighttime Flying Hours	18,160
B	Area of the Bay of Biscay (mi²)	130,000
T	Surfaced Hours per Transit	13
S	Aircraft Speed (knots)	150
W_{ASV}	ASV Mark III Sweep Width (miles)	10
W_{Naxos}	Naxos Sweep Width (miles)	30
W_{fa}	Naxos False-Alarm Sweep Width (miles)	22

The number of unforestalled ASV contacts a U-boat could expect to experience during a passage through the Bay would thus be

$$H \bullet B \bullet \frac{T \bullet F \bullet S \bullet W_{ASV}}{\sqrt{W_{ASV}^2 + W_{Naxos}^2}} = 0.47$$

Because 281 Bay transits were made during the period of Naxos use,[135] the fact that only 46 out of an expected 149 sightings were made leads to the conclusion that in 103 cases, or 37 percent, the U-boat detected the airplane through the use of Naxos. The discrepancy between 47 percent and 37 percent is not very unsettling given the large amount of estimation involved in calculating the latter figure.

The number of times per transit that a U-boat could expect to be counted as a disappearing contact is

$$\frac{T \bullet F \bullet S}{H \bullet B} \bullet \left(W_N - \frac{W_N^2}{\sqrt{W_N^2 + W_A^2}} \right)$$

or 0.23, agreeing with the actual value of 65/281. This agreement further confirms the theory.

The number of false alarms a U-boat could expect to experience during a Bay transit is found using the false-alarm sweep width W_{fa}:

$$\frac{T \bullet F \bullet S \bullet W_{fa}}{H \bullet B} = 3.2$$

Considering that a maximum-submergence order was in effect at the time, a U-boat skipper could not view with equanimity the prospect of spending an extra hour and a half surfaced, as a result of recharging after three needless crash dives. Nor would the seven-to-one (that is, 3.2:0.47) odds that a given alarm was in fact false inspire his confidence in the Naxos unit. Yet the consequences of ignoring a true attack could be severe: the skipper is on the horns of a classic threshold-setting dilemma. A *receiver operating characteristic curve* depicts such situations, showing the "probability of hit" (in this case, the probability of forestalling a radar contact or managing to disappear before the attack came) as a function of the "false-alarm rate" (in this case, false alarms per Bay transit). We will work with the somewhat more straightforward comparison of true warnings per transit to false alarms per transit. (See figure 29.) The usual procedure in such a case is to impute costs to false alarms and misses, then operate at such a point on the curve as to incur equal total costs from each source. Such a computation in the present case would lead to minimal surfacing, as Dönitz had commanded. However, such a policy raises a

Brian McCue

larger question: what amount of surfacing in the Bay is optimal from the point of view of *merchant-vessel sinking*, not U-boat survival. To answer such a question, we need to model not just the U-boats' transit of the Bay, but their whole circulation through the Bay, repair ports in France, and operating areas at sea.

Figure 29
Trade of ASV Hits and Naxos False Alarms
Effect of Varying Naxos Sweep Width

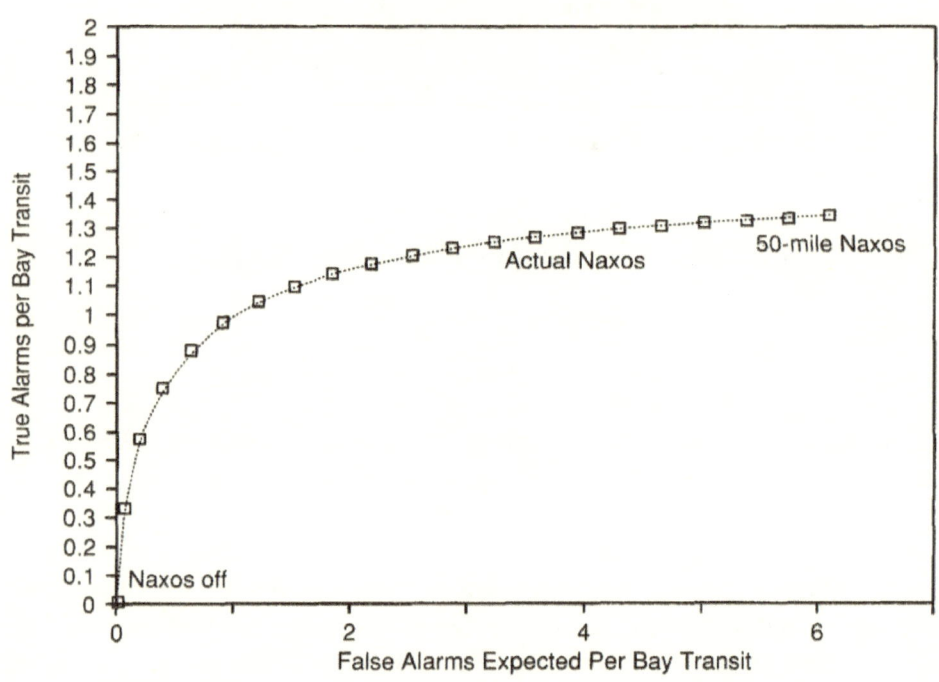

5

Morse and Kimball's
U-Boat Circulation Model

U-Boat Command regards the solution of this [U-boat repair yard] labor problem as the Navy's most important task, if we are to succeed in rapidly creating a large U-boat fleet with which to achieve great and decisive success. Its solution is also regarded as most urgent at the present juncture, when it is essential to sink the maximum possible amount of shipping this year, before the strengthening of the enemy's defenses and the expansion of his shipbuilding program begin materially to prejudice the effective-ness of U-boat warfare.

—Grand Admiral Karl Dönitz, 1942[136]

Morse and Kimball relate an anecdote about measures of effectiveness: the small number of aircraft shot down by merchant ships' antiaircraft guns did not justify their continued use, and an argument arose that the guns should be redeployed elsewhere. Recalculation using a more appropriate measure of effectiveness—merchant ships saved, not aircraft destroyed—showed that attacks against the ships with guns were far less successful than those against unarmed ships. Accordingly, the guns were retained.[137]

Similarly, the tally of U-boats sunk in the Bay does not adequately measure the effectiveness of Allied offensive search operations, though some have cited the seemingly unimpressive number of sinkings in the Bay as evidence of the inefficacy of offensive search.[138] The true measure of effectiveness is merchant ships saved, not operational sweep rate or submarines sunk. In addition to having a closer relationship to victory, the merchant-ships-saved measure of effectiveness provides a means of relating success in the Bay to success in other endeavors, such as bombing U-boat repair facilities, that call upon the same resources.

To estimate the number of merchant ships saved by the Bay offensive, we need a single model of U-boat operations that includes both Bay transit and action against merchant shipping. Although new construction added boats to the fleet and antisubmarine warfare removed them, they were intended to circulate through repair, passage through the Bay, operation against merchant shipping at sea, return through the Bay, and entry into port to be prepared for the next sortie. For this reason, wartime operations researchers called their overall model of U-boat warfare a *U-boat circulation model* even though it provided for the one-way steps of construction and destruction.

In this chapter we shall examine how Morse and Kimball assembled such a model; in the next we shall assemble a more elaborate one.

Morse and Kimball's Model

Morse and Kimball present a model created expressly to compare the effectiveness of attacks on the U-boat repair bases in France with that of attacks on the submarine-building industry in Germany. The model uses the following variables:[139]

$A =$ average number of U-boats in the Atlantic
         (including the Bay of Biscay)
$B =$ average number of U-boats in their bases
$P =$ production of U-boats per month
$S =$ sinkings of U-boats per month
$I =$ net monthly increase m U-boats t = time, in months
$L =$ U-boats leaving the bases, number per month
$M =$ maximum rate of repairing U-boats, in U-boats per month
$1/C =$ mean time, in months, to repair a U-boat in a very lightly filled base
$1/K =$ mean time spent by a U-boat in the Atlantic, in months

Two coupled differential equations connect the variables in the model:

$$\frac{dA}{dt} = P - S - AK + L$$

and

$$\frac{dB}{dt} = AK - L$$

The first equation says that the rate of increase in U-boats in the Atlantic is the net increase of the force (new production minus sinkings), minus the rate at which U-boats leave the Atlantic, plus the rate at which they leave the repair areas. The second equation says that the rate of increase in boats in the repair areas is the rate at which they return from the Atlantic minus the rate at which they are sent back out. New U-boats do not initially burden the repair capacity.

The model's basic constraint, stated by Morse and Kimball without derivation,

$$L = M(1 - e^{-CB/M})$$

governs the rate at which the repair areas ready returning U-boats for another tour of duty. (Note that the numerator and the denominator of the exponent have the same units, submarines per month.) Substituting this constraint and the identity

$$I = P - S$$

into the equations above gives

$$\frac{dA}{dt} = I - AK - M(1 - e^{-CB/M})$$

and

$$\frac{dB}{dt} = AK - M\,(1 - e^{-CB/M}).\,*$$

    While the original pair of rate equations and the identity relating sinkings and production to the inventory's rate of change are quite straightforward, the repair constraint is hardly intuitive. Although this equation and the situation bring queuing theory to mind, no standard queuing model addresses the outstanding feature of this constraint: for small values of B, throughput equals $C \cdot B$ U-boats per month, whereas "pressure" exerted by larger values of B increases throughput asymptotically to M U-boats per month.
    Morse and Kimball do not describe the genesis of this constraint. They say that the equation

> indicates that if there are a small number of submarines in the bases then the repair work can proceed efficiently enough so that the submarines can be sent out again about a month after they have come in from the previous cruise . . . . When there are a large number of submarines present in the bases, however, the state of repair of the bases and the average damage to the submarines begins to make itself felt.[140]

The variable M embodies the "state of repair of the bases," while C embodies the average damage to the submarines. Morse and Kimball used $M = 50$ and $C = 1$.

---

*   Recent attention to systems of coupled differential equations that display infinite sensitivity to initial conditions as well as m many other odd traits may cause the reader to wonder if the U-boat circulation model will suffer from such "chaotic" (the mathematical term) behavior. It will not, because such systems must consist of at least three coupled differential equations. Though some "systems" of even a single time-step equation can suffer from chaos, we will hope—with considerable justification—that the next chapter's time step version of the present chapter's differential equations model will not do so. (Gleick, p. 264; Glass & Mackey, p. 195.)

A more intuitive equation might be

$$L = \min(M, CB),$$

which says—for $C = 1$—that the repair areas repair either the maximum number of submarines they can fix in a month or the number they have on hand, whichever is fewer. This value of L would exceed Morse and Kimball's for all positive values of CB, but approximates it, especially for values of CB/M that differ greatly from unity. One could argue that Morse and Kimball really thought U-boat repair was governed by the more intuitive formula, but used theirs because it is differentiable, a requirement for their solution of the system on the Massachusetts Institute of Technology's electromechanical Differential Analyzer. One could also argue that they used their formula purely because it fits the facts well, as shown by the graph of departures per month from France as a function of U-boats in the French repair ports. (Morse and Kimball do not present such a graph and could not have presented the one below, as it shows data from the Dönitz *War Diary*.) However, either reason would represent such a departure from the authors' usual method of operation that one would expect them to call special attention to it. Tailoring an equation to suit the requirements of computers would have required considerable explanation, and the now-common practice of choosing functional forms impressionistically would have been new in 1946.

In fact, one may surmise a derivation of the constraint equation given by Morse and Kimball that is entirely consistent with their approach to many other problems.[141] Suppose that there exist M/C repair stations, each of which can repair a submarine in a month. The particular repair station to which a U-boat is assigned depends upon the nature of the damage (including normal wear and tear) it received on patrol. To minimize wasted repair capacity, the repair stations' responsibilities should be delineated so that, in the long run, they receive equal numbers of U-boats. Because damage occurs randomly, the B submarines receive, in effect, random assignments to these stations. Both M/C and B being much larger than unity, the assumptions of the Poisson distribution apply, and so the fraction of stations left without boats to repair will be about $e^{-CB/M}$, with the remainder busy and thus producing $M(1 - e^{-CB/M})$ repaired boats at the end of the month. The

Poisson nature of the assignment process can be grasped by visualizing
the submarines dropping randomly into the repair stations like buttons
dropped into a notions box: stations without any submarines in them
obviously can effect no repairs, while stations assigned one or more
boats repair just one. (See figure 30.)

This approach uses the same notion of a coverage factor and the Poisson
probability of not getting zero "hits" that we used to find the portion of
the Bay covered by lethal search; Morse and Kimball used it for several
gunfire problems[142] as well as for search problems.

Morse and Kimball worked with the model by recombining the
variables listed above into five dimensionless numbers, which they present
without the characterizations given here:

$x = CA/M$ Submarines at sea as a multiple of the repair bases'
              monthly repair capability.
$y = CB/M$ Case load of the bases: submarines in repair bases as a
              multiple of the bases' monthly repair capability.
$p = I/M$ Rate of change of force size as a multiple of the repair
              bases' monthly repair capability.
$k = K/C$ Ratio of average stay in port to average stay at sea.
$u = Ct$ Elapsed time in units of mean U-boat repair time.

With these definitions, the U-boat circulation equations become dx

$$\frac{dx}{du} = p - kx + (1 - e^{-y})$$

and

$$\frac{dy}{du} = kx - (1 - e^{-y})$$

These equations express the same related-rates relationships that the
previous pair did, only in dimensionless form. The first equation, like its
counterpart, says that the increase in the number of boats at sea is the net
production rate (production minus sinkings), minus the rate at which boats
enter repair yards, plus the repair rate. The second equation, again like

Figure 30
"Notions Box" Repair Model

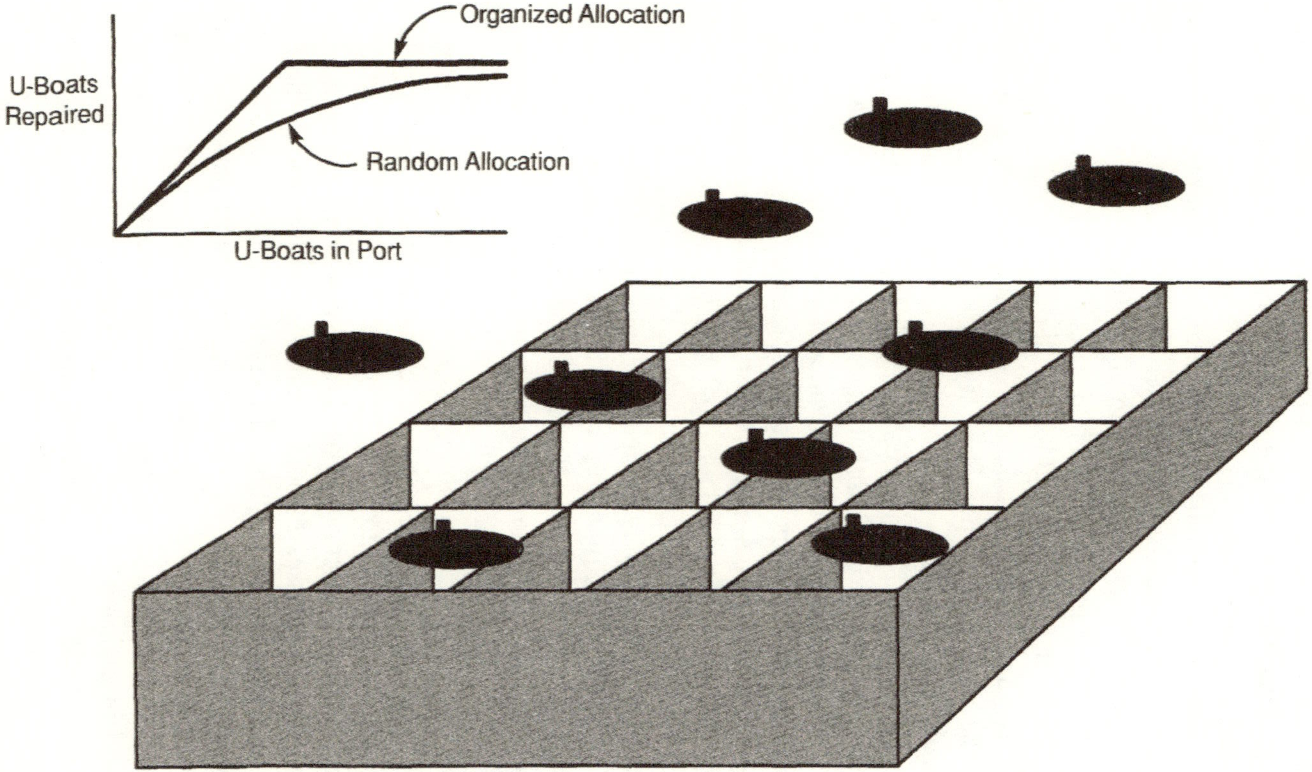

its counterpart, says that the increase in the number of boats in the yards is the rate at which boats arrive minus the rate at which they are repaired and depart. These increases, of course, could take on negative values. The equations, like the first pair, are immediate corollaries of the definitions and the repair constraint.

In an early application of automatic computing, the wartime operations researchers used the Rockefeller Differential Analyzer at MIT to solve these equations for a time span of $u = 0$ to $u = 10$, with a variety of values for the other parameters: x and y could be 0, 1, or 2, while p could be 0, 1/4, 1/2, or 1, and k could be 0, 1/2, or 1.

Regardless of the reason for switching to dimensionless numbers—be it the researchers' personal style or the demands of the Differential Analyzer—the above values of x, y, p, and k contain clues about the range of values Morse and Kimball considered plausible for the original capital-letter variables. Morse and Kimball state that $1/K = 2$ and that M ranged from 50 to 100, revealing that they considered values of up to 100 boats at sea or in port, and that the highest net rate of force increase they considered reasonable was 100 U-boats per month. Most interestingly, Morse and Kimball do not appear to have regarded a U-boat sinking rate above the U-boat production rate as a reasonable case to entertain. Perhaps this premise led to the idea of trying to use bombing to create or tighten a bottleneck in the repair yards.

Morse and Kimball show solutions that, for a wide range of the input parameters, "seem to indicate that the damaging of the repair bases had greater effect than the damaging of the factories"[143] in terms of the average number of U-boats at sea in the first few months after the attack.

The information in Dönitz's *War Diary* shows that the number of repaired boats leaving the French harbors per month exceeded 50 on only one occasion. These data also confirm the "random" repair equation advanced by Morse and Kimball; the curve—based on a maximum repair capacity of 50 boats per month—passes through month—passes through the cloud of points representing actual monthly numbers of boats in port and boats repaired in 1942 and 1943. (See figure 31.) In contrast, the straight line—based on servicing all boats in port, not to exceed 50—generally lies above the cloud.

Morse and Kimball highlight repairs because their model was to be used in making decisions about whether to bomb the repair areas. The rest of their U-boat circulation model is sketchy: for example, sinkings do not depend on the number of boats at sea. Such considerations call for creating a new U-boat circulation model such as will be developed in the next chapter.

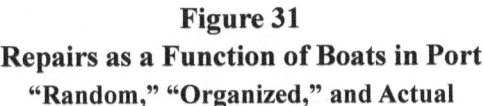

Figure 31
Repairs as a Function of Boats in Port
"Random," "Organized," and Actual

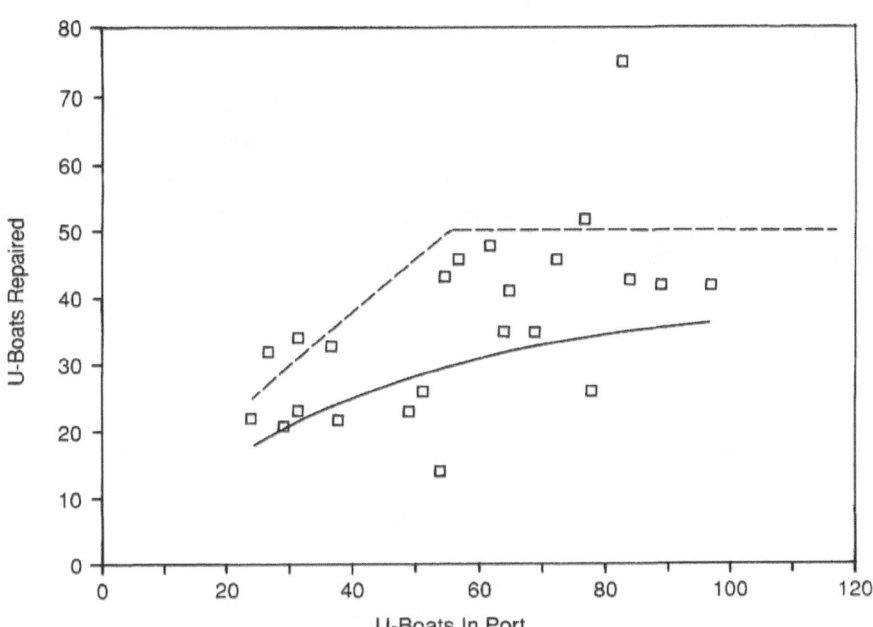

Morse and Kimball's Measure of Effectiveness

Morse and Kimball used the average number of U-boats in the Atlantic as their measure of effectiveness; they modeled damage to either the repair bases or the construction yards by lessening either M or P, then examined the result in terms of the value of A after the passage of some time. One could argue that they really should have examined the integral of A—total U-boat months spent at sea—over that interval of time, because the damage done by the U-boats would be roughly proportional to the number of U-boat months spent at sea. Perhaps the Differential Analyzer could not perform the necessary double integration.

Measuring Offensive Search Effectiveness With Morse and Kimball's U-Boat Circulation Model

Morse and Kimball do not separate U-boat losses by cause; in particular, they do not distinguish losses inflicted by convoy escorts from those inflicted by offensive search. Yet we know that U-boat tankers counteracted Allied offensive search in the Bay by making passage though the Bay unnecessary; one would wish for a way to investigate this effect quantitatively, especially if faced by a three-way choice among the use of airplanes for the two bombing tasks (against repair areas and production areas) and the offensive search mission. As we have seen, Morse and Kimball addressed this issue using estimates of the number of merchant ships saved per 1,000 hours of flying for the three missions.[144] Explicit representation of offensive search in the U-boat circulation model rapidly leads to a different sort of circulation model altogether, addressed in the next chapter.

6

A More Modern U-Boat Circulation Model

There remained, however, two points at which individual [U-boats] were brought together within a limited area, and so presented a concentration of targets which invited attack. The first was the area of transit to and from port, particularly the Bay of Biscay. The second was the refueling [rendezvouses] on which most cruises to distant areas were dependent. These vulnerable points offered the opening through which the Allies drove home their successful [antisubmarine] offensive in the summer and fall of 1943.

—Anonymous author of National Security
Agency document SRH-008, 1945[145]

We will now set up a U-boat circulation model according to a paradigm more familiar to modern readers than that of Morse and Kimball, namely that of multiple barriers.[146]

The Model

Figure 32 shows the full structure of the model. U-boats originate in German shipyards and proceed to patrol areas in the North Atlantic where, with older boats operating out of occupied France, they seek to

Figure 32
U-Boat Circulation Model

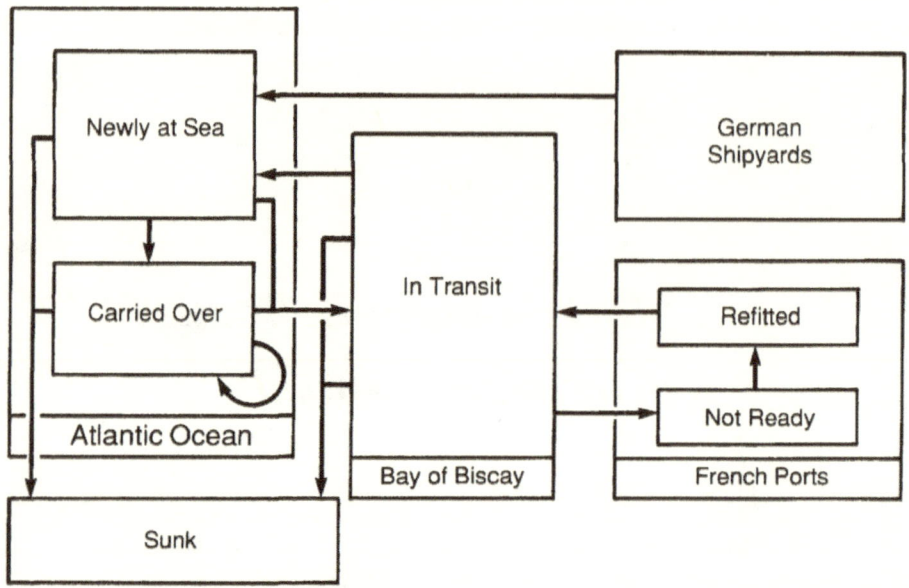

destroy merchant shipping. Boats running low on fuel and other supplies may either go into the French ports or take on fuel at sea from a tanker submarine. Boats in the Atlantic run a risk of being sunk during their attacks on convoys; boats returning to France run an additional risk when passing through the Bay of Biscay. Boats arriving in France undergo repair and become available for another unsafe trip through the Bay and back to the hunting grounds in the North Atlantic.

The model contains simplifications, but they are not—considering its purposes—oversimplifications. For example, the model rules out the return of any boats to Germany, though in fact operational boats returned to Germany on a handful of occasions throughout 1942 and 1943.[147] It disregards the operation of ports in Norway, on the grounds that passage to and from Norway involved so few boats. Similarly, the model makes no explicit allowance for the passage of boats to the Southern Hemisphere and the Indian Ocean. For generality's sake the model allows a refueled boat to be refueled at sea repeatedly: the Germans did refuel a certain boat twice in one voyage,[148] and the exclusion of such refueling on the grounds that it rarely happened might introduce an artificial constraint into "what-if" analyses.

This U-boat circulation model steps through time in discrete one-month increments. Each month some of the submarines in French ports leave, some of the ones at sea return, and some new submarines set out from German ports. Submarines at sea find and sink merchant ships. Patrols in the Bay of Biscay and the defenses of convoys act as filters (other authors use the term "barriers" in connection with this type of model) to U-boat passage, sinking some passing U-boats and allowing the rest to go through.

Sinkings in the Bay depend on the amount of Allied search effort there, the effort's effectiveness, expressed as a sweep rate, and the policy pursued by the U-boats. For now, to aid in understanding this attrition model, we will assume a 10 percent chance that a U-boat is sighted during each Bay transit and a 33 percent chance of being sunk if sighted. Thus the U-boat has a 3 percent chance of not surviving a trip through the Bay, for a complementary 97 percent chance of surviving a one-way transit.

Sinkings in the Atlantic obviously depended on the amount of time spent on patrol, the frequency of contacts with convoys, and the hazard of such contacts to the U-boats. The model requires estimates of these parameters.

Transiting the Bay of Biscay would take at least one day, and as many as five under a policy of maximum submergence.[149] The U-boat would spend up to 12 days in continued transit beyond the Bay to the patrol area.[150] Time at sea thus greatly exceeded time on station as counted in Dönitz's *War Diary*, though the Diary notes that U-boats often found targets while in transit from or to the desired station.

During the period in question there were likely to be between seven and eight Allied convoys present at any time in the 3-million-square-mile region where the U-boats lay in wait.[151] Thus—as a round figure—one may assume 0.0000025 convoys per square mile. A U-boat could maintain an operational sweep rate (as defined previously in the case of aircraft searching for submarines) of at least 2,500 square miles per day against these convoys. With five submarines in a wolf pack searching at this rate with no significant overlap (a reasonable assumption given the way the submarines operated), the pack could expect to encounter about 0.03 convoys per day.

Given a 5 percent chance per engagement that a U-boat is sunk by a convoy and a 0.03 percent chance per day that it meets a convoy, the product becomes the daily probability of an encounter fatal to the U-boat, and the complementary probability becomes that of surviving a day of patrol. Because the days may be considered statistically independent, the probability of surviving 10 days on station is the 10th power of the probability of surviving a single day: with the values assumed here, between 98 and 99 percent:

$$(1 - 0.03 \cdot 0.05)^{10} = 0.985.$$

Because the voyage's patrol period and its two transit periods are all mathematically independent, the probability that the submarine survives the whole cruise is the product of the probabilities of surviving the individual parts: two transits and 20 days in the patrol area. In our illustrative case, the submarine has a 93 percent chance of surviving the mission, because

$$0.97 \cdot 0.985 \cdot 0.97 = 0.93.$$

The complementary probability of 7 percent is the chance of not surviving. Dividing that into 1 (taking the reciprocal) gives the life expectancy of the U-boat—14 voyages—if none of the above probabilities changes.* The probabilities did, as we know, change often; variation in the dangers presented by the different barriers resulted in different steady-state life expectancies for the U-boats. The "steady-state life expectancy" wraps up the barriers' effects at any moment into a convenient measure of effectiveness. Figure 33 shows steady-state U-boat life expectancy measured in months at sea during the 1942-43 period.

The foregoing illustrates the attrition portion of a U-boat circulation model: we also seek a sense of the boats' "productivity" during their careers. Some sense of that productivity comes from considering only the time spent at sea, not in refit or in waiting for refit.

---

*   The life expectancy is the expected number of voyages given the probability p of sinking during any one voyage. The underlying distribution is the geometric distribution, so the expected value is $1/p$.

Figure 33
Steady-State U-Boat Life Expectancy
Measured In Months at Sea

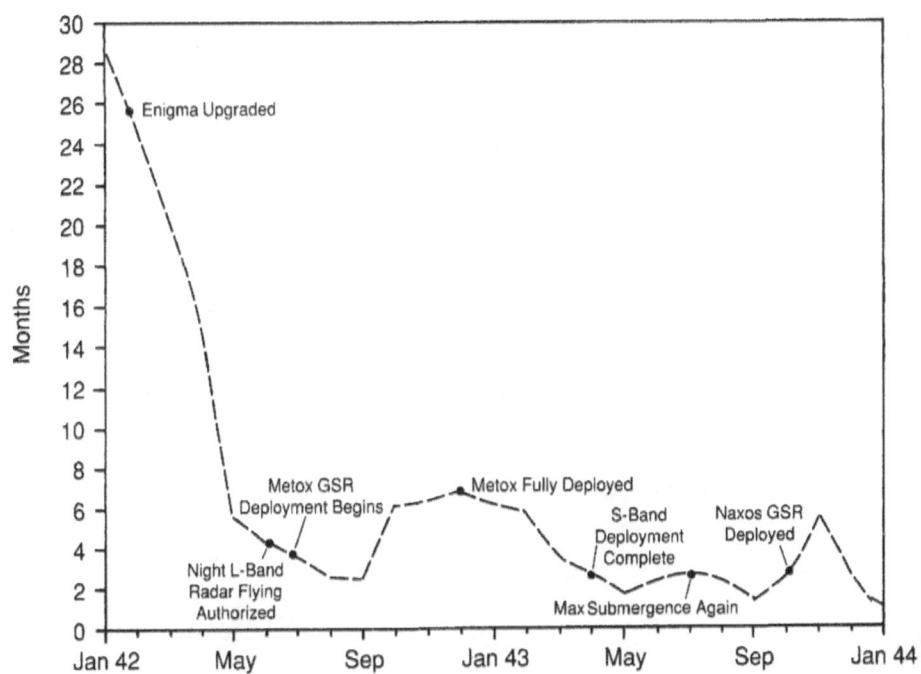

Given that convoys lost about one merchant vessel per attacking submarine in wolf pack attacks,[152] and given the convoy encounter rate computed above, in the particular case at hand the U-boat could expect to find seven-and-a-half convoys in its career, sinking roughly one ship in each.

Finally, having addressed U-boat attrition and productivity, we need an estimate of the repair capacity of the U-boat ports. For *M*, the maximum number of boats that could be repaired in a month, Morse and Kimball used 50, a value that fits the repair data gleaned from Dönitz's *War Diary*. However, the *Diary* also indicates that some boats operating out of the French ports were not assigned to attack merchant shipping in the North Atlantic. Some patrolled the South Atlantic, some the Indian Ocean, some were assigned "special missions" such as dropping off spies in America, and some were U-boat tankers.[153] Thus the repair rate for the submarines whose circulation we seek to model must be reduced—we will use a value of 40 as the maximum number of boats repaired per month.

The next several sections will provide the mathematical details of the circulation model outlined above. The reader who finds these details onerous or obvious may skip to the section titled "Measuring Effectiveness in the U-Boat Circulation Model."

Details Regarding New U-Boats Coming From Germany

New boats come from Germany; in each month i a number of new boats NEW_i set out. The historical production figures[154] used here may appear low to the reader conversant with German submarine production, but the figures are for *combat-ready* U-boats entering service in the North Atlantic, not for the launching of all types of boats intended for all theaters.

Details Regarding U-Boats in the Atlantic

The model portrays the Atlantic as populated by two groups of U-boats: those carried over from the previous month and those not. The carryover can occur, as explained above, through refueling or simply through frugality. For want of a better term, the boats not carried over from the previous month will be called "nonrefueled." Thus, for month i,

$$ATSEA_i = NONREFUELED_i + CARRYOVER_i$$

New boats, represented by the variable NEW_i, arrive automatically in the Atlantic without attrition on the way because they are coming from Germany, not passing through the Bay of Biscay. (See table 14.) Experienced boats surviving the Bay passage join the new boats to form the nonrefueled contingent. These experienced boats are considered to have made the Bay passage in the previous month, so we have

$$NONREFUELED_i = NEW_i + OUT_{i-1}.$$

The defining formula for OUT_i will be presented below.

Table 14.—Calculated Numbers of U-Boats at Sea and Total Sinkings

		New from	Boats Out				
i	Month	Germany	Nonrefueled	Carryover	Total	Tankers	Sunk
1	Jan 42	19	21	21	42	0	1.43
2	Feb	13	43	10	53	0	2.09
3	Mar	14	39	13	52	0	2.54
4	Apr	9	44	12	56	1	3.57
5	May	4	40	23	63	1	5.15
6	Jun	13	35	25	59	2	5.09
7	Jul	24	44	34	77	3	7.65
8	Aug	32	52	47	100	2	9.71
9	Sep	32	58	42	100	2	9.73
10	Oct	27	64	43	106	3	11.60
11	Nov	11	61	54	115	3	13.79
12	Dec	14	46	55	101	2	11.90
13	Jan 43	14	50	42	93	3	13.79
14	Feb	25	51	50	101	3	11.80
15	Mar	13	61	52	113	4	21.35
16	Apr	22	49	63	112	3	19.17
17	May	16	56	53	109	4	23.41
18	Jun	7	46	61	107	1	14.76
19	Jul	3	31	33	65	0	6.59
20	Aug	2	38	15	52	0	4.38
21	Sep	18	34	12	46	2	3.13
22	Oct	12	52	31	83	0	8.43
23	Nov	17	43	19	62	0	2.58
24	Dec	9	51	15	66	0	5.83
25	Jan 44	7	43	15	58	0	5.25

Boats could stay at sea for somewhat more than a month. To stay at sea much longer, boats required refueling by U-boat tankers, which could provide about 10 "tankees" with another full load of fuel and supplies.[155] As a first-order approximation we may treat the carryover of nonrefueled boats by saying that some proportion p of boats at sea will remain at sea for another month even without refueling. (The historical data suggest that p is about 25 percent.[156]) Thus if TANKERS$_i$ tankers at sea in month i can refuel some number TANKEES of U-boats, then

$$\text{CARRYOVER}_{i+1} = \text{TANKEES} \cdot \text{TANKERS}_i + p \cdot \text{ATSEA}_i$$

except when there are not enough submarines at sea to take on all the fuel available. Strictly speaking, therefore,

$\text{CARRYOVER}_{i+1} =$
$\qquad \min(\text{TANKEES} \cdot \text{TANKERS}_i + p \cdot \text{ATSEA}_{i-1}, \text{ATSEA}_i - \text{SUNK}_i)$

where SUNK_i, the defining formula for which appears below, is the number of boats sunk in the Atlantic (as opposed to in the Bay) in month i.

U-boats neither sunk in the Atlantic in month i nor destined for the refueled contingent in the following month, $i+1$, must head for the French ports, so

$$\text{INBOUND}_i = \text{ATSEA}_i - \text{CARRYOVER}_{i+1} - \text{SUNK}_i$$

which is not to say that all the inbound U-boats will actually survive passage through the Bay. Note that the definition of NON-REFUELED_i means that a boat that was outbound in month i -1 will, if it does not get refueled, be inbound in month i: in other words, a boat spending a month at sea does not count as leaving and returning in the same month.

So far the equations presented have really expressed only definitions, matters of bookkeeping, and immediate results of the assumption that a U-boat could spend somewhat more than a month at sea without resupply and another month if refueled. Now we must begin to invoke some facts, and some of the kinds of reasoning we have seen in earlier chapters.

Once in the Atlantic, U-boats ran little risk when attacking independent merchant shipping, but ran considerable risk in attacking convoys. As we have seen, the number of U-boats lost to convoy escorts per merchant vessel sunk climbed steadily from 0.2 to 0.8 during the period of interest. Additionally, U-boats at sea faced about a 2 percent chance each month of sinking as a result of accident or striking a mine.[157] Thus we may form the relation

$$\text{SUNK}_i = (0.2 + 0.6 \cdot i/24) \cdot \text{MVSUNK}_i + .02 \cdot \text{ATSEA}_i$$

with $i = 1$ in January 1942, and MVSUNK_i being the number of convoyed merchant vessels sunk in month i.

Details Regarding Sinking of Convoyed Merchant Vessels

We must now determine MVSUNK_i. Here again the concept of sweep rate arises, but with U-boats sweeping for convoys rather than airplanes sweeping for U-boats as in the Bay of Biscay. Intelligence available to the U-boats, the lethality of U-boat attacks, and the respective densities of U-boats and convoys must also figure in the calculation.

Sternhell and Thorndike cite a 9.4-mile U-boat operational sweep width for individual merchant ships in darkness;[158] this width comes from U.S. submarine experience against Japanese ships. Sternhell and Thorndike state that because the convoy's dimensions—unlike that of a single ship—are non-negligible, the nighttime operational sweep width for convoys will be about half again as large: we will use 15 miles. They also cite a 14.5-mile width for daytime search but point out that, although ships' power plants emitted smoke only intermittently, a sizable convoy would have at least one ship emitting smoke at almost any given time, so the true sweep width of a U-boat for convoys by daylight would be about 25 miles, the distance at which the smoke could be seen. These widths, multiplied by the submarine's surfaced speed (10 knots) and by a "dynamic enhancement"* figure equal in this case to 120 percent,[159] yield daytime and nighttime sweep rates of about 300 and 170 square miles per hour for each U-boat. These rates, prorated according to the hours of daylight and darkness in each month, result in an overall sweep rate for the month.

U-boats did not search the Atlantic in total ignorance. The Allies' ciphers for transmitting information regarding convoys were often broken by the Germans. Immediately after the war, operations analysts[160] quantified the German ability to exploit such information in Operations Evaluation Group (OEG) Study 533.

Starting with the basic identity

$$\frac{\text{convoys sighted}}{(\text{convoys/square mile}) \cdot \text{U-Boat days}} = \text{U-boat sweep rate}$$

the authors of OEG Study 533 calculated the sweep rate in square miles per day (see table 15), based on known values for the quantities on the

---

* The convoys' motion makes them easier to find: with moving convoys even a stationary U-boat would have a nonzero sweep rate. This effect, quantified at length by Koopman, was not significant in the Bay of Biscay calculations because even surfaced submarines can be considered stationary with respect to 150-knot airplanes.

left-hand side. Overall, the sweep rate during periods of cipher compromise was about double that during periods of no compromise. The reader may well object that the no-compromise rate is far less than the average rate calculated above. The rates calculated in OEG Study 533 are lower because the report limits the meaning of "contact" to "first contact," not repeat contacts against the same convoy later in its voyage, saying "contact" thus "must not be confused with 'sighting' or 'detection.'" Given the purposes of the report, such a definition of contact and the resulting rate makes sense: we will use the report's conclusion about the utility of radio intercept while retaining our own definition of sweep rate, which properly includes repeat detections because these are at least as likely to produce sunken merchant vessels as are first contacts.

Table 15.—Summary of OEG Study 533 Findings

Convoy Status	Sighted	Convoy Density (convoys per mi.²)	U-Boat Days	Sweep Rate (mi.²/day)
Compromised	23	0.9/3,000,000	18,666	3,950
Uncompromised	227	5.7/3,000,000	18,666	1,900

This treatment, characteristic of the across-the-board use of sweep rates prevalent at the time, may strike the modern reader as a bizarre and formalistic use of the search-rate methodology. A modern analyst, concluding that decryption doubled the U-boats' ability to find convoys, would be likely to invoke the "force-multiplier" concept and say that the German subs enjoyed an effective doubling of their number, not of their range of vision.

Mathematically, of course, either interpretation balances the equation. Heuristically, however, one might more easily accept a third interpretation: that the convoy density was effectively increased because the U-boats could rule out, based on their radio intercepts, some fraction of the area. In this interpretation, the cipher compromise doubled U-boat effectiveness by providing the U-boat force with a priori information as to which half of the search region the convoy occupied.

This interpretation leads directly to an information-theoretic way to think about the U-boats' gains from decryption of radio messages about convoys. A "bit" is the amount of information needed to express one dichotomous distinction.[161] Though often thought of as a 0 or a 1, a bit can be transmitted in any way: one lamp or two, for example, served to inform Paul Revere whether the British were to leave Boston by land or by sea. Thus, where the authors of OEG Study 533 saw a doubling of the sweep rate, an information theorist would see the acquisition by the U-boat force of one bit of information, reducing the search area to a certain 1,500,000 of the original 3,000,000 square miles when the object of the search was a compromised convoy.

Although the above interpretation best expresses the OEG findings in terms of today's concept of information, the easiest way to compute sightings, given a varying mix of compromised and uncompromised convoys, is to regard the doubled effectiveness against compromised convoys as a doubling of their number, rather than as a doubling of the submarines' range of vision, a doubling of the submarine force, or a halving of the search area. Thus

$$\text{SIGHTINGS}_i =$$
$$(\text{PATROL}_i \cdot \text{SWEEP}_i \cdot (\text{CONVOYS}_i) \cdot (1 + \text{COMP}_i/\text{AREA})$$

where PATROL_i is the number of U-boats searching for convoys, SWEEP_i is the sweep rate of the U-boats in month i, CONVOYS_i is the average number of convoys at sea in month i, COMP_i is the proportion of convoys compromised by radio intercepts, and AREA is the 3-million-square-mile area searched.

Mathematically,

$$\text{SWEEP}_i = 10 \cdot 1.2 \cdot (25 \cdot \text{DAYLIGHT}_i + 15 \cdot \text{NIGHTTIME}_i)$$

reflecting the 10-knot speed, the 20-percent dynamic enhancement, daytime and nighttime search rates as explained above, and the hours of daylight and nighttime in month i.

The average number of U-boats actively engaged in attacking convoys during month i, PATROL_i, is less than ATSEA_i because many boats—often a majority, according to Dönitz's *War Diary*—will, at any given time, be making passage to or from the patrol area. Our U-boat circulation model has to distinguish the time spent crossing the Bay, BAYTIME_i, from the time spent passing between the Bay and the patrol area, TRANSIT, because it will later be combined with the Bay search model in which time in the Bay depends upon degree of submergence. These times are both expressed in hours: BAYTIME_i is subscripted while TRANSIT is not because the former can change but the latter will be left constant.

How many boats are actually on patrol? We find out by subtracting BAYTIME_i + TRANSIT hours from the time U-boats could be searching for convoyed merchant ships. Newly built boats, boats newly at sea from the Biscay ports, and those boats that will stay out for the coming month (either because they refuel or because they belong in the fraction p that stay out without refueling) don't incur this penalty. Many boats will belong in two of those categories, having gone to sea and returned as soon as possible: they incur the penalty twice. Thus

$$\text{PATROL}_i = \text{ATSEA}_i - \frac{(\text{TRANSIT} + \text{BAYTIME}_i)}{\text{MONTH}_i} \cdot \text{NEW}_{i-1} + \text{OUT}_i + \text{ATSEA}_i - \text{TANKERS}_i \cdot \text{TANKEES} - p \cdot \text{ATSEA}_i$$

We want to estimate the number of merchant vessels sunk, not the number of convoys contacted, so we need an additional constant L (for "lethality") by which to multiply the number of sightings to get the number of sinkings:

$$\text{SINKINGS}_i = \frac{L \cdot \text{PATROL}_i \cdot \text{SWEEP}_i \cdot \text{CONVOYS}_i (1 + \text{COMP}_i)}{\text{AREA}}$$

This equation models the U-boats' search for convoys as a clean sweep, a point to which we shall later return. The sample calculation earlier in the chapter used a value of unity for L based on wartime records of merchant vessels sunk per attacking U-boat.* This value is probably too high for L in the present equation: it should be degraded to reflect the fact that not all U-boat sightings of convoys resulted in attacks,[162] and because the value is based on "wolfpack attacks"[163] to the apparent exclusion of any attacks by single U-boats. The data clearly show that larger wolfpacks sank disproportionately more merchant vessels, so we may conclude that attacks by unaccompanied boats would lower the average value of merchant vessels sunk per attacking boat. Other data not specifically cited as based on wolfpack attacks point to fewer than one merchant vessel sunk per attacking U-boat.[164] Taking all these considerations into account, we may assign a value of 0.8 to L, the number of merchant vessels sunk per attacking U-boat.

---

*   Any single value of L is an oversimplification, because L depended on the number of attacking U-boats. Lanchester theorists might investigate Morse & Kimball, pp. 46-47, and Sternhell & Thorndike, pp. 106-109.

Table 16.—Calculated Sinkings of Merchant Vessels and U-Boats at Sea (Sinkings of Independent Merchant Vessels Explained Below)

i Month	Convoy Density	% Compromised	Convoyed M/V Sunk	Indep M/V Sunk	Sunk by Escorts	Other Sinkings
1 Jan 42	8	0	3.0	52	0.60	0.83
2 Feb	8	0	4.4	72	1.03	1.06
3 Mar	8	0	5.7	76	1.50	1.04
4 Apr	8	0	8.3	88	2.45	1.12
5 May	8	0	11.8	102	3.88	1.27
6 Jun	8	0	10.9	93	3.91	1.19
7 Jul	7.6	3	15.6	99	6.10	1.55
8 Aug	7.6	3	18.2	113	7.72	1.99
9 Sep	7.6	3	16.9	91	7.72	2.01
10 Oct	7.6	3	19.4	75	9.47	2.13
11 Nov	7.6	3	22.1	63	11.49	2.30
12 Dec	7.6	3	17.9	44	9.87	2.03
13 Jan 43	7.1	27	20.4	39	11.94	1.85
14 Feb	7.1	27	15.9	46	9.79	2.01
15 Mar	7.1	27	29.5	57	19.08	2.27
16 Apr	7.1	27	24.9	58	16.93	2.24
17 May	7.1	27	29.8	59	21.23	2.18
18 Jun	5.6	17	16.9	55	12.61	2.15
19 Jul	5.6	17	6.8	25	5.30	1.29
20 Aug	5.6	17	4.1	17	3.34	1.04
21 Sep	5.6	17	2.6	12	2.21	0.92
22 Oct	5.6	17	7.8	17	6.77	1.65
23 Nov	5.6	17	1.5	10	1.35	1.24
24 Dec	5.6	17	4.8	8	4.52	1.32
25 Jan 44	5.6	17	4.2	7	4.09	1.16

Details Regarding Passage Through the Bay of Biscay

Working from the definition of the kill sweep rate, we may find the ratio of U-boats sunk in the Bay to their average density

$$\left(\frac{\text{Sinkings}}{\text{Density}}\right)_i = P(K|S)_i \cdot \text{SWEEP}_i \cdot \text{HOURS}_i$$

where SWEEP_i is the operational sweep rate of aircraft patrolling the Bay. (See table 17.) We will want to exercise the model with sweep rate values derived from our sweep rate model as well as

Table 17.—Calculated U-Boat Sinkings in the Bay of Biscay

i month	Five-Month Moving AV P(K|S)	Sinkings Density	Inbound Boats Transits	Inbound Boats Sinkings	Outbound Boats Transits	Outbound Boats Sinkings
1 Jan 42	0%	0.00	30	0.00	21	0.00
2 Feb	0%	0.00	38	0.00	24	0.00
3 Mar	0%	0.00	37	0.00	26	0.00
4 Apr	0%	0.00	29	0.00	30	0.00
5 May	2%	0.07	34	0.13	31	0.12
6 Jun	3%	0.13	21	0.15	31	0.23
7 Jul	3%	0.15	22	0.57	32	0.80
8 Aug	4%	0.29	47	1.30	29	0.79
9 Sep	4%	0.28	48	1.33	27	0.74
10 Oct	2%	0.04	41	0.16	32	0.13
11 Nov	2%	0.03	46	0.16	34	0.12
12 Dec	2%	0.03	47	0.12	35	0.09
13 Jan 43	1%	0.02	29	0.05	36	0.06
14 Feb	1%	0.05	37	0.18	37	0.19
15 Mar	3%	0.11	29	0.30	37	0.37
16 Apr	4%	0.21	39	0.83	37	0.77
17 May	7%	0.59	24	1.36	36	2.01
18 Jun	21%	1.70	60	9.96	36	6.04
19 Jul	22%	1.74	44	12.75	35	10.13
20 Aug	21%	0.26	36	1.58	36	1.60
21 Sep	25%	0.58	12	1.23	36	3.59
22 Oct	23%	0.27	56	2.52	36	1.61
23 Nov	12%	0.10	44	0.76	32	0.54
24 Dec	12%	0.21	45	1.55	35	1.21
25 Jan 44	16%	0.32	40	2.11	36	1.92

with the historical values. This basic Bay equation relates to the other equations via variables denoting the numbers of submarines attempting passage of the Bay in each direction, $OUTBOUND_i$ and $INBOUND_i$:

$$INSUNK_i = \left(\frac{Sinkings}{Density}\right)_i \bullet INBOUND_i \bullet (BAYTIME_i/MONTH_i)$$

$$OUTSUNK_i = \left(\frac{Sinkings}{Density}\right)_i \bullet OUTBOUND_i \bullet (BAYTIME_i/MONTH_i)$$

Surviving boats exit the Bay:

$$ARRIVED_i = INBOUND_i - INSUNK_i$$

$$OUT_i = OUTBOUND_i - OUTSUNK_i.$$

Details Regarding Repair of U-Boats in Biscay Ports

We have already seen, at the end of Chapter 5, the "notions box" model of repair activities. In terms of the variables we have been using,

$$OUTBOUND_i = REFITTED_{i-1}$$

$$NOTREADY_i = NOTREADY_{i-1} + ARRIVED_i - REFITTED_i$$

and

$$REFITTED_i = MAXREPAIR \bullet (1 - e^{-NOTREADY_{i-1} / MAXREPAIR})$$

See table 18, which shows monthly calculations of REFITTED; based on arrivals and the resulting total number of boats in port, as well as newly produced boats and the resulting number of "nonrefueled" boats at sea.

**Table 18.—Calculated Repair Activity in French Ports
(New Boats Entering Atlantic From Germany and Nonrefueled
Total Shown for Reference Purposes)**

i Month	Arrivals	Total in Port	Boats Refitted	New from Germany	Nonrefueled
1 Jan 42	30	37	24	19	21
2 Feb	38	43	26	13	43
3 Mar	37	55	30	14	39
4 Apr	29	62	31	9	44
5 May	34	60	31	4	40
6 Jun	21	62	32	13	35
7 Jul	22	51	29	24	44
8 Aug	46	44	27	32	52
9 Sep	47	63	32	32	58
10 Oct	41	78	34	27	64
11 Nov	46	85	35	11	61
12 Dec	47	95	36	14	46
13 Jan 43	29	106	37	14	50
14 Feb	36	98	37	25	51
15 Mar	29	98	37	13	61
16 Apr	39	90	36	22	49
17 May	23	93	36	16	56
18 Jun	50	80	35	7	46
19 Jul	31	95	36	3	31
20 Aug	34	89	36	2	38
21 Sep	11	88	36	18	34
22 Oct	53	63	32	12	52
23 Nov	44	85	35	17	43
24 Dec	43	93	36	9	51
25 Jan 44	38	100	37	7	43

Measuring Effectiveness in the U-Boat Circulation Model

Although operational sweep rate made a good measure of effectiveness for judging the effort of Allied patrol aircraft in the Bay, measuring the effectiveness of the U-boat force is more difficult. One possible measure of merit, used by Sternhell and Thorndike, is merchant vessels sunk per submarine built. This measure matters because Germany could ill afford industrial competition with the United States, and thus needed to inflict the most damage per submarine.

As we do not intend to investigate alternative uses to which the Germans might have put the effort and resources consumed in U-boat production, we may take that production rate as given and use merchant ships sunk as a measure of the U-boat force's effectiveness.

Data for the U-Boat Circulation Model

The U-boat circulation model uses monthly figures for

1) the part of U-boat production that was destined for the North Atlantic,
2) the density of convoyed shipping in the North Atlantic,
3) the amount of information about convoy routing available to the Germans,
4) the activities of U-boat tankers,

as well as the previously reviewed monthly data on flying hours, sweep rates, and probability-of-kill-given-sighting. Some constants appear as well, most of which have already been mentioned: the areas of the regions treated as "the North Atlantic" and "the Bay of Biscay," the transit time—exclusive of time spent in the Bay—between the Bay of Biscay and the U-boats' operational area, the U-boats' daytime and nighttime sweep rates when searching for convoys, the maximum rate at which the ports could ready returned U-boats for their next cruises, the number of U-boats replenished by a single tanker, and the number of U-boats available at the beginning of 1942 (see table 19).

OEG Study 533 treats three periods of U-boat warfare: July 1942-December 1942, January 1943-May 1943, and September 1943-March 1944. In those periods decryption of Allied ciphers compromised, respectively, 3 percent, 27 percent, and 17 percent of the convoys. Sternhell and Thorndike state that a U-boat making contact could expect to sink an average of eight-tenths of a merchant vessel.[165]

Rossler provides a detailed accounting of U-boat tanker activities, reporting each instance of refueling and noting which U-boats received the fuel. For our purposes, since we are assuming that each tanker refuels exactly 10 boats, we may extract from Rossler's account the occasions of refueling and then count each as servicing 10 U-boats. Dönitz's *War Diary* provides a month-by-month count of the U-boats entering North Atlantic service.

Table 19.—Important Constants of the Bay Search Model

3,000,000	Atlantic Shipping Area (sq. mile)
130,000	Bay of Biscay Area (sq. mile)
12	Transit Time (days) Exclusive of Bay Transit
312	Daytime U-Boat Sweep Rate for Convoys (sq. mile/hour)
176	Nighttime U-Boat Sweep Rate for Convoys (sq. mile/hour)
40	Maximum Economic Repair Rate (boats/month)
10	Number of Boats Refueled per Tanker
78	Initial Total Number of U-Boats
42	Initial Number of U-Boats at Sea, Evenly Divided Between Unrefueled and Carried Over
0.25	Proportion of Unrefueled U-Boats Carrying Over Anyway
0.8	Merchant Vessels Sunk per U-Boat per Convoy Sighting

To compare the model's performance with reality, we need monthly figures for U-boats at sea and convoyed merchant vessels sunk. These can be readily obtained[166] and are best viewed graphically along with the results of running the model.

Comparison of the Model to Reality

To make sure that nothing is terribly wrong with the model, we will first use it to "predict" the damage done to convoyed merchant shipping, given the historical data. (See figure 34.) The U-boat circulation model satisfactorily tracks the historical numbers of U-boats at sea and convoyed merchant ships sunk. (Morison does not list sinkings of convoyed merchant ships in months when fewer than four were sunk, hence the gaps in the series.)

Interestingly, the August 1943 "retrenchment," often seen as a command decision, follows purely as a consequence of our assumptions about refueling and U-boat endurance, and does not require any additional variable "policy" embodying Dönitz's decisionmaking.

Figure 34
U-Boats at Sea and Convoy Sinkings

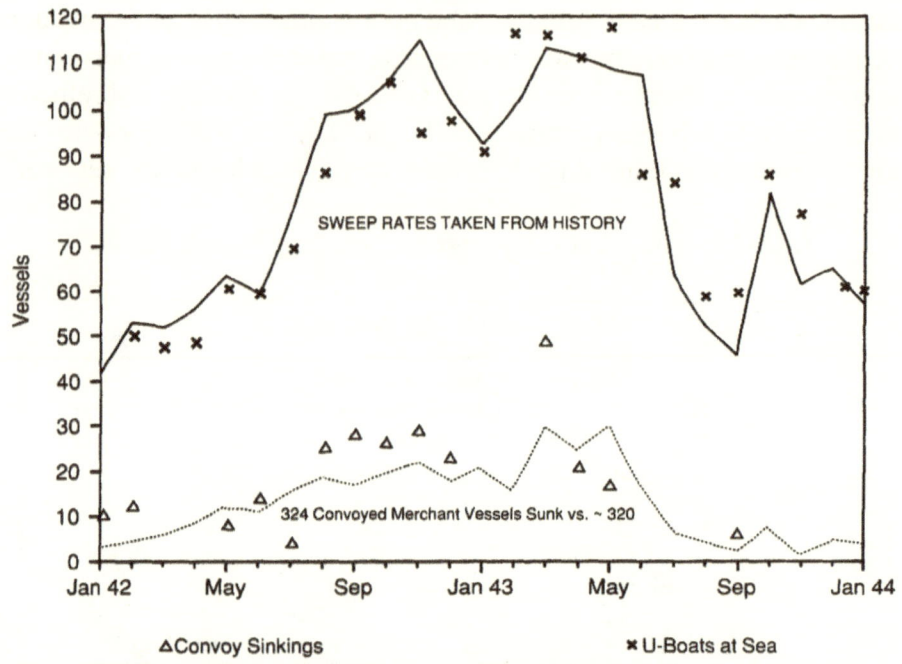

△ Convoy Sinkings   ✗ U-Boats at Sea
● *Line Equals Calculated Values; Symbols Equal Actual Values.*

The model also tracks historical losses, though not as well: some fluctuation around the values returned by the model is to be expected—not only because of random departures from what amount to expected-value calculations, but also because the convoy densities and percentages of compromise are averages over periods of several months, and will thus result in smoother-than-historical outcomes. Satisfied that the model works, we will return and add one last equation by which to estimate the losses of non-convoyed merchant vessels.

Losses of Independent Merchant Shipping

So far our efforts have focused on the U-boats' ability to sink convoyed shipping, and although the model's agreement with reality in this regard validates the modeling process, the model will not be useful unless it also estimates losses of independent merchant shipping, which in fact predominated in 1942 and 1943.

Unlike the convoys, which deliberately zig-zagged and used other means to make the expected value of their density as uniform as possible in the U-boats' region of operation, independent shipping tended to frequent certain areas. Hence intelligence—if only that communicated from U-boat skipper to U-boat skipper—probably played an even greater role here than it did in the attack on convoyed shipping. A model of independent merchant-vessel attrition could easily become more complicated than the entire U-boat circulation model developed so far, and would require far more data. Therefore we will take a totally different approach: multiple regression to fit an impressionistically derived functional form. This approach offers an instructive contrast to the method favored thus far, in which the functional forms are the fruits of the wartime operations researchers' work and the parameter values are a matter of record, leaving no undetermined coefficients free to be chosen purely for the sake of good fit.

We will assume that the sinking of independent ships occurred in daylight and in direct proportion to their number and that of the U-boats at sea, and that the number of independent ships at sea declined exponentially during the course of the war as convoying became more and more prevalent. These assumptions make sense because encounters with independent shipping occurred randomly—and thus in proportion to their density—and because of the decisiveness with which the convoy system was introduced.

The first assumption restates the equation used earlier for sinkings of convoyed merchant ships:

$$S(t) = k \cdot D(t) \cdot U(t) \cdot M(t)$$

where $D(t)$ is the daylit fraction of the day, $S(t)$, $U(t)$, and $M(t)$ are the numbers of sinkings, U-boats at sea, and merchant ships at sea—all functions of time—with the constant of proportionality k embodying everything else about speed, sweep width, proportion of time spent on station, information, and so on, all treated as unchanging.

The second assumption gives an equation for $M(t)$:

$$M(t) = M(0)e^{-Kt}$$

where $M(0)$ is the number of independent merchant ships at sea at the outset (December 1941) and K is a factor expressing the rate at which independent shipping declined.

The equations combine, forming

$$S(t) = k \cdot D(t) \cdot U(t) \cdot M(0)e^{-Kt}$$

which can be made linear by taking logarithms

$$\ln(S(t)) = \ln(k) + \ln(D(t)) + \ln(U(t)) + \ln(M(0)) - Kt$$

and rearranged to form

$$\ln(S(t)) - \ln(U(t)) - \ln(D(t)) = \{\ln(k) + \ln(M(0))\} - Kt,$$

an equation whose left side consists of known quantities and whose right side consists of the constant in curly brackets minus a term proportional to the passage of time. We will fit a straight line to the transformed data and then transform the line back to the exponential expression we seek.

Least-squares regression gives values of 1.433 and 0.09086 for $\{\ln(k) + \ln(M(0))\}$ and K, respectively. The antilogarithm of 1.433 is 4.192, so we have

$$S(t) = 4.19 \cdot D(t) \cdot U(t) \cdot e^{-.0909t}.$$

As figure 35 shows, this equation passes a curve through the data in a convincing way.*

Not only is the above derivation unlike any other so far, it leads to a totally different type of result.** The number 1.433 equals $\{\ln(k) + \ln(M(0))\}$,

---

*   More technically, 78 percent of the variance in sinkings is "explained," with 23 out of 25 degrees of freedom remaining.

**   Certain parameter values estimated heretofore are notable for not being regression results. The number of "tankees" per tanker and the probability p that an unrefueled U-boat stays at sea an extra month could have been determined by regression methods, regressing returns per month on U-boats and tankers at sea, but they were not. Instead they were separately estimated from the exact history of refuelings, which Rössler provides, and the endurance of submarines. The 25 percent chance that a U-boat stays out for a second month is really an expression of a 1.33-month endurance, laid on the Procrustean bed of a model that operates in one-month steps.

Figure 35
U-Boats at Sea and Independent Sinkings

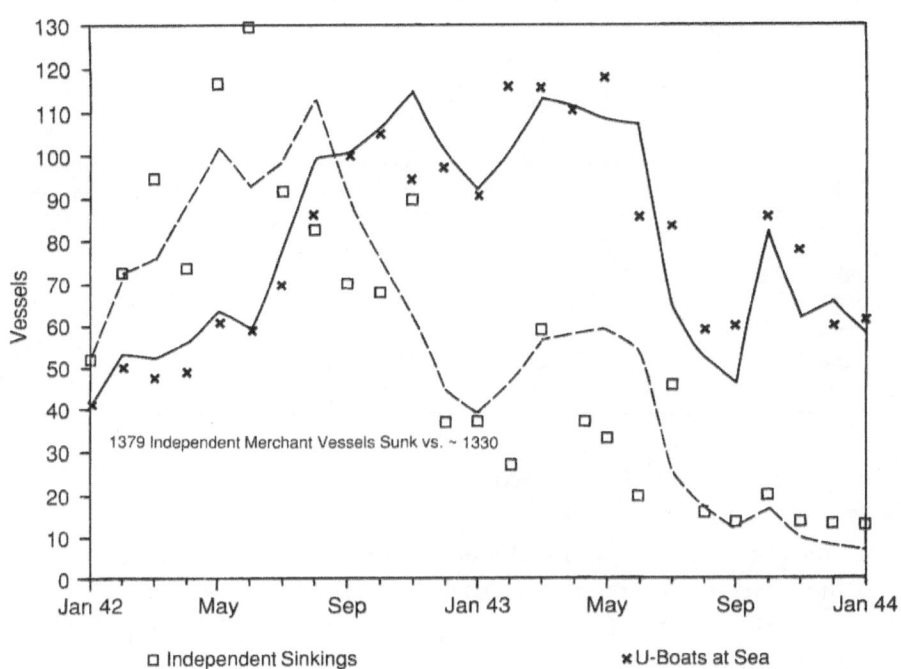

● *Line Equals Calculated Values; Symbols Equal Actual Values.*

but nothing indicates how much of 1.433 is the logarithm of the catch-all constant k or how much is the logarithm of the original number of independent merchant ships. Despite statisticians' appropriation of the verb "to explain" as a technical term, the regression analysis provides no explanation at all.

Indeed, the entire line of reasoning is after-the-fact. Our previous modeling efforts used hardware characteristics (such as speed and sighting distance) and tactics (such as submergence) to arrive at calculated values irrespective of their historical counterparts, and caused us to think through the processes we sought to model. In contrast, curve-fitting methods work backward from the values they purport to predict, arriving at coefficients (and, in some cases, even a whole model[167]) justified solely on the grounds of a good fit.

This good fit can result from compensating errors. The U-boat circulation model has a substantially better fit (in terms of reducing the sum of the squared departures from the actual values of U-boats at sea, convoyed merchant vessels sunk, and independent merchant vessels sunk) if one uses the curve-fitter's values presented in table 20 instead of the fact-based values shown in table 19.

Table 20.—Contrast Between Curve-Fitting and Use of Facts

Parameter	Curve-Fitter's Value		Fact-Based Value
Daytime U-Boat Sweep Rate	245	sq mi/hr	312
Night U-Boat Sweep Rate	305	sq mi/hr	176
Maximum U-Boat Repair Rate	58	boats/month	40
"Tankees"/Tanker	5	boats/tanker	10
Percent Carryover Without Refueling	27	(pure number)	25
Convoyed Vessels Sunk/Sighting	0.9	(pure number)	0.8

The departures in the last two variables are not disturbing, but serious distortions appear in the other four. The repair rate is far higher than the Germans could attain (and reflects no use of Biscay port capacity for repairing boats not used in the North Atlantic) and the number of "tankees" is—in compensation—far too low. Even worse, the boats appear more capable of finding convoys at night than in the daytime, which we can tell is absurd by reading accounts of convoy battles, such as those in SRH-008. Yet the fit is good.

I highlight this point brightly because the use of regression analysis has become so prevalent that some experienced analysts—when first exposed to the idea of working *from* radar and receiver sweep widths, the forestalling equation, airplane speed, and so on *toward* a sweep rate—assumed that the entire exercise consisted of regressing, or even guessing, a coefficient for each piece of equipment so as to get as close a fit as possible to the historical data. Interestingly, graduate students taking a course in which some of the present material was developed had far less trouble than experienced analysts.

Exercising the Circulation Model

Now that we have seen that the model reproduces historical data tolerably well, we are in a position to run it under different assumptions about Bay search, with the results made manifest in the number of merchant vessels sunk. Somewhat surprisingly, the best way to do so is by comparing the results of different runs of the model with each other, rather than with historical outcomes.*

The most obvious question to ask is "How much good did Bay search operations do?" By running the model with and without Bay search operations, we find that a seemingly small number of ships were saved by the search: 26, or fewer than 2 percent. This finding is surprising—we said above that a U-boat could expect to sink about seven-and-a-half merchant vessels in its career; why does the model (in which more than 80 U-boats are sunk in the Bay) credit the Bay flying effort with so few merchant vessels saved?

The answer lies in the maintenance backlog. A U-boat would have to be very lucky to receive repeatedly the rapid turnaround time implicitly supposed above.

But each ship carried enough supplies to support an infantry division in Europe for over a week! A fighting man consumes about 100 pounds of supplies per day, and a division contains about 10,000 men.[168] The lost ships' capacities average over 5,000 gross registered tons (CRT).[169] Equating 5,000 CRT to 10 million pounds,** we find that a shipload of supplies could last a 10,000-man division 10 days.

---

*   Alan S. Blinder, in his macroeconomic analysis of the U.S. economy in the 1970's, asks many questions about the effects of President Nixon's price controls. Such inquiry about effects contains, as Blinder explicitly notes, the counterfactual inquiry "What if there had been no constraints?" Blinder answers these questions by creating a time-step model of the U.S. economy, running it with and without controls, and attributing the difference to the controls. He suggests that the model-model comparison, unlike the more straightforward model-reality comparison, will net out any systematic error in the model and will also avoid artifactual attribution of month-to-month random fluctuation to Nixon's price controls. We shall take up the whole issue of counterfactual questions shortly.

**   An approximation implicit in the definition, but an underestimate for most cargoes: a gross registered ton is defined as 100 cubic feet of carrying capacity. Most ships can in fact carry more tons-weight than gross registered tons. In any case, any cargo denser than balsa wood weighs more than 20 pounds per cubic foot. (See also Alden, pp. xxv-xxvi.)

The calculation that Bay search saved only 26 merchant ships is naive in that it assumes that the Germans would operate in the same way regardless of whether Allied search aircraft patrolled the Bay. If we assume that, in the absence of any Bay threat, U-boats would have crossed the Bay on the surface, we find that the extra time the U-boats could have devoted to hunting merchant shipping would have made a greater difference than did the mere lack of U-boat sinkings in the Bay: 145 more merchant ships would be sunk than in the base case. For this reason, we may characterize Bay search as quite effective: a few thousand flying hours saved more than enough cargo to supply a division of Allied soldiers on the Continent from D-Day until V-E Day—almost a year. Very roughly, a man-day of flying effort (even including ground-crew efforts) saved 4 man-years of supplies.

"What if . . . ?"

Historians normally abhor the "What if . . . ?" questions, such as those implicit in the previous section. David Hackett Fischer points out the logical difficulty such "counterfactual conditionals" entail: "All of the historical 'evidence' for what might have happened if Booth had missed his mark is necessarily taken from the world in which he hit it."[170] Historians often do address "Why?" questions, answering that one or more events caused another, but—as Fischer also points out—such statements beg the question of what would have happened if one or more causes had not been present.[171]

Economists and political scientists show less reluctance in this regard, perhaps because they recognize that policymakers who ask "Why?" will not accept a philosophical refutation of causality as an answer. Bernard Brodie, for example, feels free to state that "while it would have proved very difficult to defeat the U-boat without the aircraft, it would have been impossible to do so without the escort vessel."[172] Herbert York, writing on the American decision to build the H-bomb, explicitly posited a counterfactual world in which President Truman decided, in January 1950, to cancel the weapon.[173] York's methodology fails to disentangle him from the trap enunciated by Fischer.

Some Alternative Courses of Action and Their Consequences

The U-boat circulation model provides grounds for asking and answering what-if questions which turn into empirical questions about the

Figure 36
Campaign Without Convoy Escorts

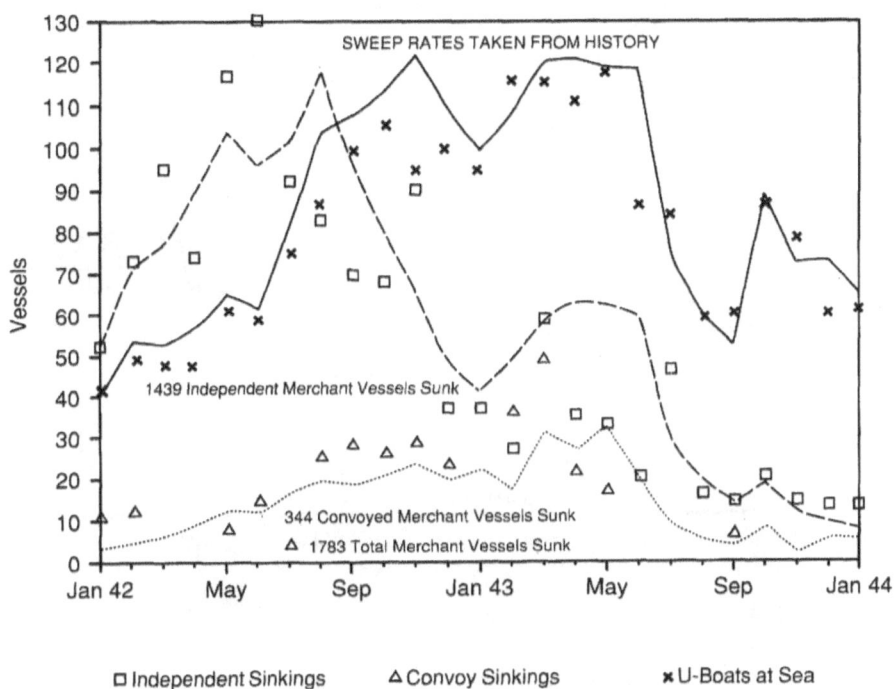

SWEEP RATES TAKEN FROM HISTORY

1439 Independent Merchant Vessels Sunk

344 Convoyed Merchant Vessels Sunk

1783 Total Merchant Vessels Sunk

□ Independent Sinkings   △ Convoy Sinkings   × U-Boats at Sea

● *Line Equals Calculated Values; Symbols Equal Actual Values.*

model's behavior. The model can measure the effectiveness of the tactics and countermeasures used as well as some that were not used. The basis for comparison will be the difference (in merchant vessels sunk) between the case at hand and the model's "base case," the number of merchant vessels it tallies as sunk when presented with the conditions that actually prevailed in the Second World War.

We may logically begin with Brodie's implicit counterfactual questions: "What if there had been no antisubmarine aircraft?" and "What if there had been no antisubmarine surface escorts?" Let us assume that surface escorts accounted for *all* the sinkings of U-boats by convoy escorts. (This assumption doubtless overstates the contribution of the surface craft; airplanes escorted convoys as well. Yet merely adjusting the U-boat/merchant-ship exchange rate to reflect the rate of U-boat sinkings by aircraft would understate the ships' contribution, as the two types of escort

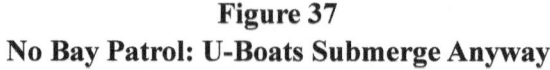

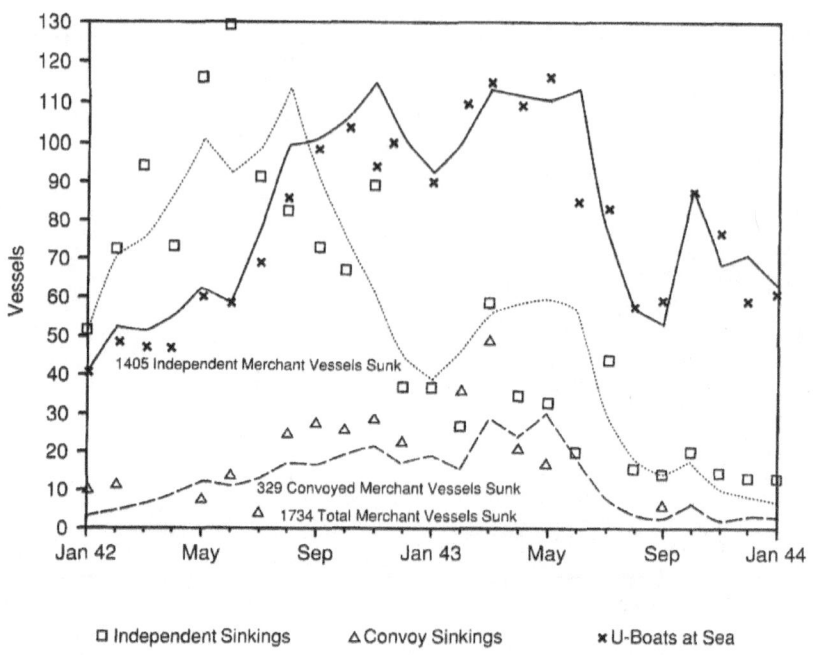

Figure 37
No Bay Patrol: U-Boats Submerge Anyway

☐ Independent Sinkings   △ Convoy Sinkings   ✕ U-Boats at Sea

● *Line Equals Calculated Values; Symbols Equal Actual Values.*

cooperated.) Then we can assess the relative contributions of surface vessels and airplanes by comparing a hypothetical campaign without convoy escorts to the one in which aircraft did not patrol the Bay. Running the model under these two assumptions (see figures 36, 37, and 38), we find that the U-boats are more successful in the no-aircraft case than in the no-escort case. The reason is an effect noted earlier, but that Brodie had probably not taken into account: in the no-airplanes case, the U-boats' time at sea increases because they are free to transit the Bay on the surface. Without the Allied use of aircraft, the U-boats could sink 145 more merchant ships than in the base case; if the Allies had used aircraft but no surface escorts, the U-boats would have sunk only 80 extra ships.

The Germans recognized the value of extended time at sea. Refueling a U-boat already at sea would permit it to remain in operation another month without the time-consuming and increasingly dangerous trips to

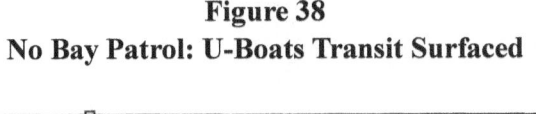

Figure 38
No Bay Patrol: U-Boats Transit Surfaced

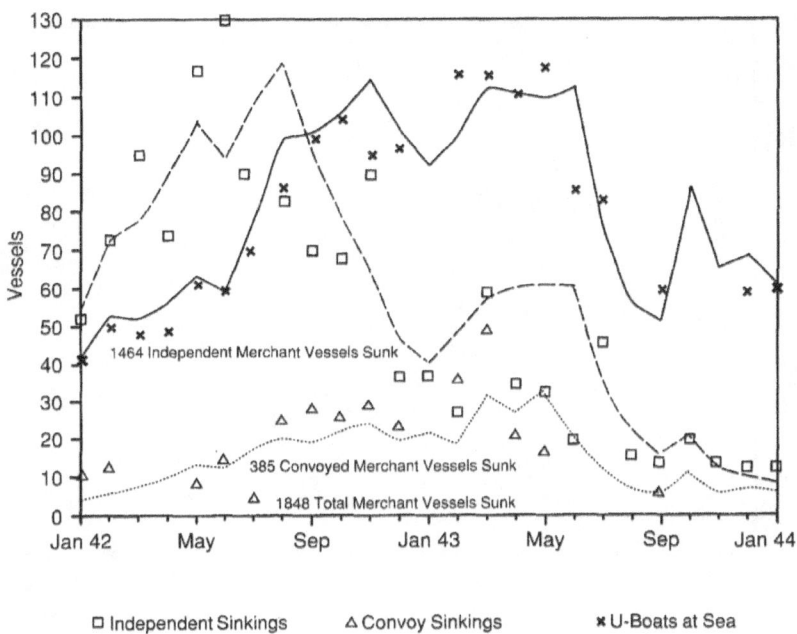

port and back again. Yet conventional submarine tenders would never be able to sneak out to sea or survive there. Special tanker submarines appeared as early as March 1942;[174] later, replenishment U-boats appeared that could resupply combat U-boats with torpedoes and even fresh-baked bread. The U-boat circulation model illustrates the value of the U-boat tankers. Although 1,650 merchant ships were actually sunk, the model calculates 1,703: it is to this value that we must compare the model's calculated result for a war fought without tanker U-boats (see figure 39): only 1,219 merchant ships in the period considered. With the number of U-boat tankers doubled from 11 to 22, the Germans would have, according to the model, sunk 2,050 merchant ships. (See figure 40.)

Such wide swings in effect require, of course, a *ceteris paribus* caveat: we have assumed that "all other things remain equal." But there is no particular reason to think that they would not have remained equal without

Figure 39
Effect of Zero U-Boat Tankers

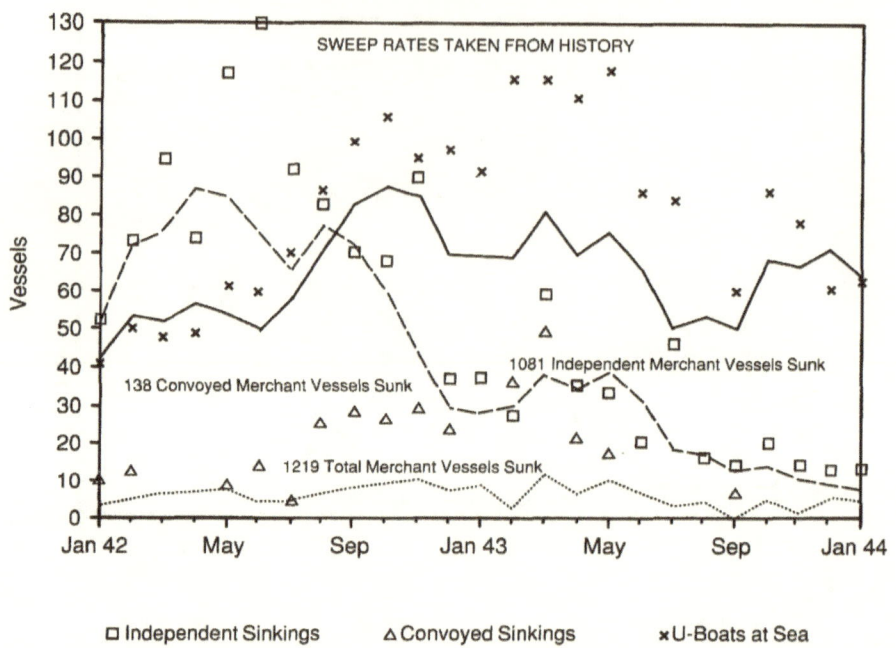

□ Independent Sinkings   △ Convoyed Sinkings   ×U-Boats at Sea

● *Line Equals Calculated Values; Symbols Equal Actual Values.*

tankers: the model provides a believable assessment that the tanker U-boats caused the loss of over 500 merchant ships. Eleven extra tanker boats could easily have been built (in fact, ten were started and then scrapped[175]), and the means of their defeat—Enigma decryption—already received top priority from the Allies and so could not have been spurred to greater achievements by a greater possible payoff.

Characterizing the U-boats' search for convoys as a clean sweep is justified because of the U-boats7 close cooperation and Dönitz's up-to-the-hour coordination. We could alter the equation to reflect a random search as opposed to a clean sweep, modeling a hypothetical campaign in which the U-boats did not use wolfpack tactics. In that case, reasoning analogous to that used in characterizing Bay search as random gives

$$\text{SINKINGS}_i = \text{CONVOYS}_i (1 - e^{-L(\text{PATROL}_i \cdot \text{SWEEP}_i \cdot (1 + \text{COMP}_i)/\text{AREA})})$$

Figure 40
Effect of Doubled U-Boat Tankers

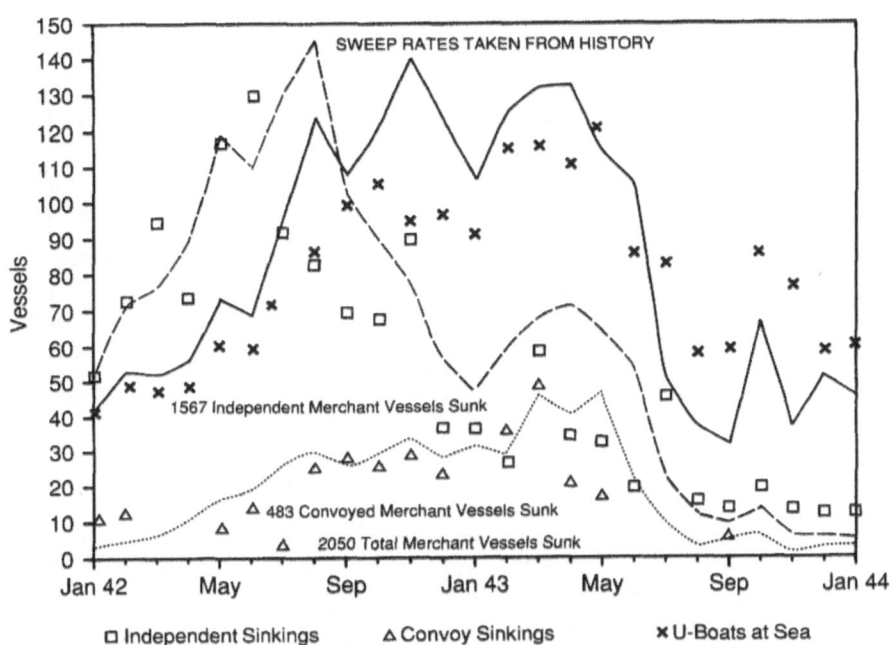

● *Line Equals Calculated Values; Symbols Equal Actual Values.*

This equation can be used to show that the U-boats would have sunk far fewer convoyed ships had they searched independently. (See figure 41.) The U-boats probably also benefited from the coordinated attacks made possible by wolfpacks: U-boats would move along with a convoy until several were present, then attack it all at once. Thus the random-search equation deprives the U-boats of only part of the benefit of wolf-packs, and any comparison based on it will understate the true advantage conferred by the wolf pack concept of operations. One may note also that by finding fewer convoys, the U-boats lessen their own losses and thus slightly increase the losses of independent shipping.

We may also run the model with data depicting the use of much better submarines, which the Germans had but never actually used. (See figure 42.) These advanced U-boats, such as the Type XXI, featured a snorkel and a hull form optimized for underwater performance rather than surface

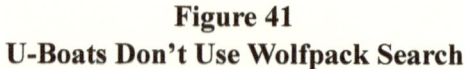

Figure 41
U-Boats Don't Use Wolfpack Search

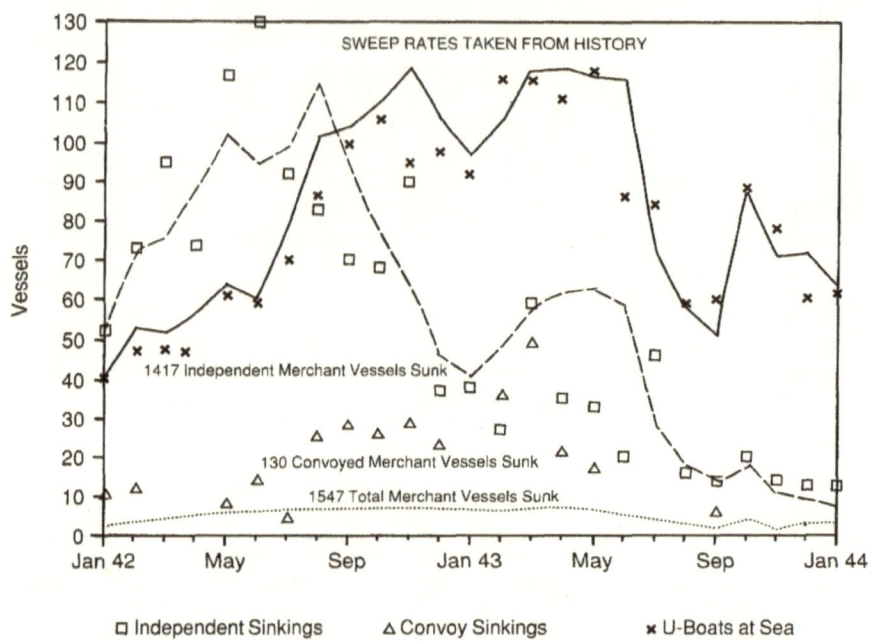

□ Independent Sinkings   △ Convoy Sinkings   ✗ U-Boats at Sea

● *Line Equals Calculated Values; Symbols Equal Actual Values.*

performance: their dirigible-shaped hull differed from previous U-boats' boat-shaped hull, which had a distinct bow, deck, and guardrails. In fact, even though 119 Type XXIs were delivered, only a few went to sea, too late in the war to have much effect.[176]

The Type XXI was ahead of its time even late in the war (the United States did not build such a submarine until a decade later), but we may well wonder how the U-boat campaign would have gone if Germany had used these true submarines, rather than mere "undersea boats," throughout the war.* The Type XXI's high submerged speed would make it much harder

---

*   Indeed, Soviet Admiral S. G. Gorshkov speculates counter factually on the possible results of German use of "new types of submarine," doubtless including the Type XXI and the various Walter boats. (The Sea Power of the State, p. 265.)

Figure 42
Hypothetical Type XXI Campaign

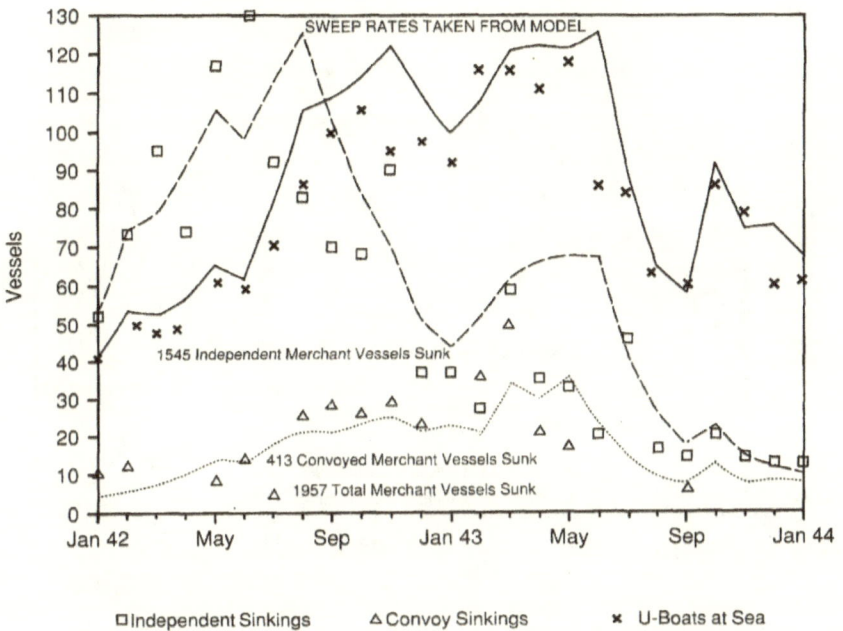

Independent Sinkings   △ Convoy Sinkings   × U-Boats at Sea

● *Line Equals Calculated Values; Symbols Equal Actual Values.*

to destroy during attacks on convoys; it would also have a much shorter, and thus safer, Bay transit time. Taking those advantages into account, by eliminating all attrition except for accidents, we can see that the Type XXI adds about 250 merchant sinkings to the total—a sizable increase. Arguably, the true increase would be greater, because the Type XXI's other technical advances might well make it more deadly in attacks on convoys, raising the number of merchant vessels sunk per convoy sighted.

U-boats awaiting repair accumulated in France. Because they were in an occupied country, repair facilities could not be expanded easily, and extra manpower was not forthcoming from Germany. (The facilities could, however, be made very damage resistant—Allied bombing had little effect on the famous concrete submarine pens.) We may use the U-boat circulation model to explore the effects of improved U-boat maintenance by doubling the French ports' repair capability. (See figure 43.)

Figure 43
Doubled Repair Capacity

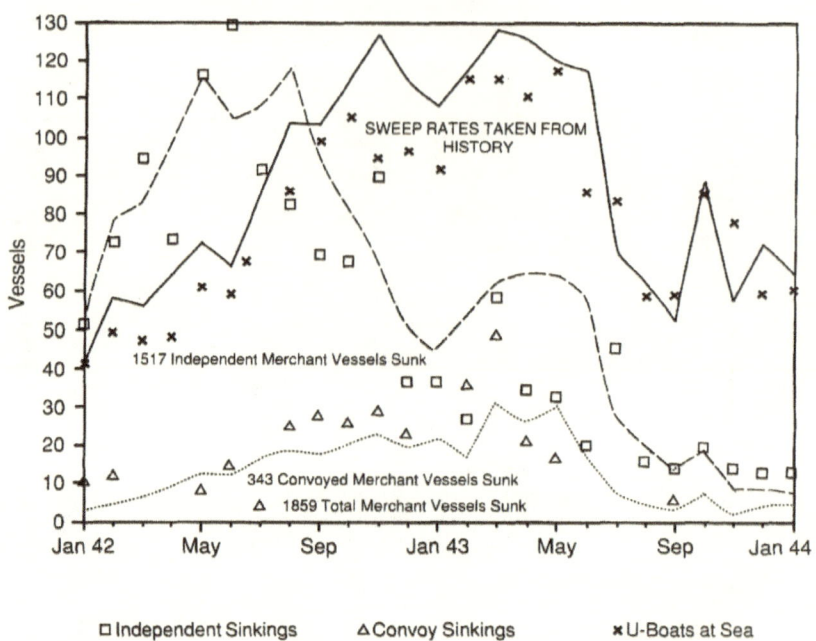

● *Line Equals Calculated Values; Symbols Equal Actual Values.*

Some authors[177] imply that U-boats retrenched in August 1943, in reaction to the ASV Mark III radar. Dönitz certainly called back some boats that had already sailed.[178] However, the model displays the visible signs of retrenchment without any special programming to do so: a major decline in U-boats at sea results purely from the policy of maximum submergence and the paucity of at-sea refueling capability by that time. We can use the model to assess the effect of an earlier retrenchment, starting in May 1943: the ASV Mark III went into service in April and Dönitz certainly could have implemented his retrenchment in May instead of August. (See figure 44.) Allied shipping could have operated unmolested, but if German submarine production continued unabated, Dönitz could count on unleashing an enormous fleet of U-boats once their survivability could again be ensured. Although a modeled retrenchment in May saves some U-boats, it actually lessens the number of merchant vessels sunk because of the ever-declining opportunity to sink independent merchant ships.

Figure 44
Early Retrenchment Campaign

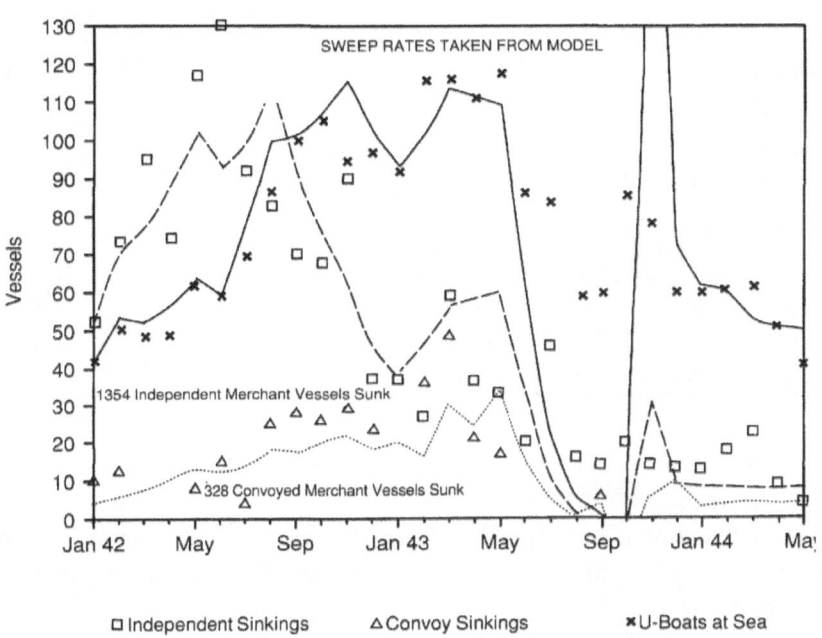

● *Line Equals Calculated Values; Symbols Equal Actual Values.*

Conclusions From the U-Boat Circulation Model

Table 21 summarizes this chapter's findings. It shows the merchant vessels sunk for the cases considered here, the actual history, and the model's base case—the result it calculates based on historical data. The absence from the base case of any retrenchment causes a few extra sinkings, but some of the discrepancy between the base case and the actual history doubtless arises from mistaken parameter estimates. However, we will not revise these parameters in a curve-fitting attempt to make the base case match the actual history. The sensitivity of the model to changes in inputs (our whole reason for using it) would also be altered in such an adjustment, and we have no reason to think that this sensitivity would be altered for the better. Therefore we will retain the best-estimate parameter values we have been using and use the model's base case, not the actual history, as our point of departure.

Table 21.—Results of Excursions Using the U-Boat Circulation Model

Case	Merchant Vessels Sunk	
	total	vice base
Actual History	1650	− 53
No U-Boat Tankers	1219	−484
No Wolfpack Search	1547	−156
Early Retrenchment	1682	− 21
Model Base Case	1703	0
No Convoy Escorts	1783	+ 80
No Bay Patrol (U-boats don't submerge)	1848	+145
Doubled Repair Capacity	1859	+156
Hypothetical Type XXI Campaign	1957	+254
Doubled U-Boat Tankers	2050	+347

The model shows that additional U-boat tankers would have caused the most dramatic change, that the convoy escorts saved 80 merchant ships, that the Bay patrol saved 145 ships, and that doubled repair capacity (probably an impossibility) would have sunk more than 150 extra merchant ships.

These conclusions raise other questions about the specifics of Bay operations: what if the Germans had used maximum submergence the whole time, or minimum submergence, or had been quicker to realize which electronic countermeasures to employ? To answer these questions, we must couple the U-boat circulation model developed in this chapter with the model of Bay search operations, developed in Chapter 4, to form an omnibus model.

7

The Omnibus Model

While it is convenient to examine the many aspects of search in themselves and in some degree of isolation, it would be fatally misleading if, in so doing, one were to fall into the trap of those specialists who, in developing one part of a whole, ignore the rest and its relation thereto. Splendid work has been done in medical research by detailed examination of one disease or one physical function, but such results must be integrated into the whole human system if they are to serve effectively.

—Bernard Osgood Koopman, 1980[179]

The "omnibus model" combines the Bay search model with the U-boat circulation model, allowing investigation of the effects of changing Bay search conditions on the U-boat campaign as a whole. Using such a model, we can assess the effectiveness of alternative Bay policies by their results on merchant-shipping losses as measured in the model.

Creating the Omnibus Model

The basis for comparing policies is the same as in the previous chapter: the difference (in merchant vessels sunk) between the case at hand and the model's "base case." (See figure 45.) The slight difference between historical sweep rates and those computed by the sweep-rate model results in a difference of four merchant ships: in the previous chapter's model the

Figure 45
Omnibus Model: Base Case

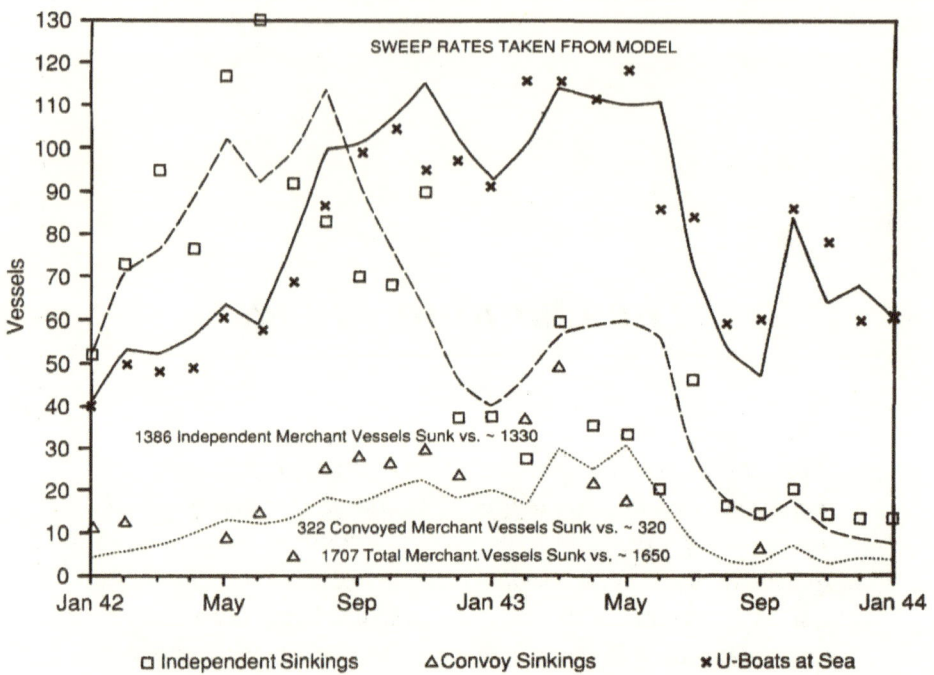

SWEEP RATES TAKEN FROM MODEL

1386 Independent Merchant Vessels Sunk vs. ~ 1330

322 Convoyed Merchant Vessels Sunk vs. ~ 320

1707 Total Merchant Vessels Sunk vs. ~ 1650

□ Independent Sinkings   △ Convoy Sinkings   ✕ U-Boats at Sea

● *Line Equals Calculated Values; Symbols Equal Actual Values.*

historical sweep rates resulted in a base case of 1,703 ships sunk, whereas the omnibus model—using computed sweep rates derived by the methods of Chapter 4—computes a base of 1,707 merchant ships sunk.

Exercising the Omnibus Model

As mentioned earlier, Dönitz withdrew his boats from the Atlantic in the summer of 1943 because he had lost all his U-boat tankers. Any such "retrenchment" when the S-band threat was recognized, to wait until a search receiver was ready, would have been counterproductive because of the declining density of independent shipping. To assess the full effect of retrenchment, we must run the model—with and without early retrenchment (see figures 46 and 47)—through May 1944 to give the Germans the full

Figure 46
Campaign Through D-Day

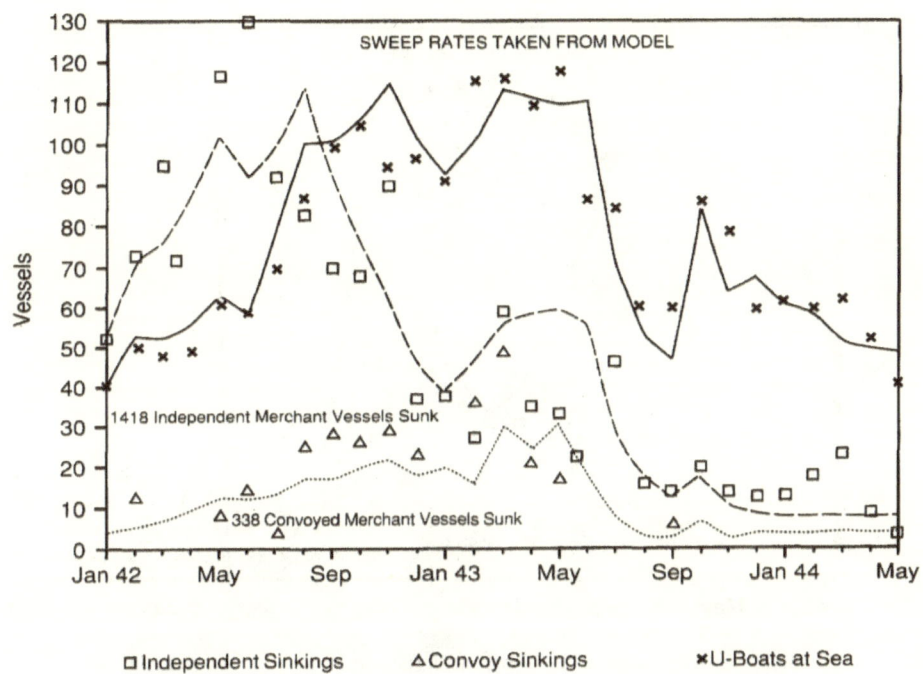

□ Independent Sinkings △ Convoy Sinkings ✕ U-Boats at Sea

● *Line Equals Calculated Values; Symbols Equal Actual Values.*

benefit of their huge U-boat fleet in the Atlantic until they turn it against cross-channel shipping after the Normandy invasion. Using calculated sweep rates in the omnibus U-boat circulation model (rather than the historical ones, which are available only through January 1944) lets us simulate the snorkel portion of the U-boat campaign in order to assess the full impact of the retrenchment option.

Again, the retrenchment option has hardly any effect on the total number of merchant ships sunk. Given that continued U-boat activity tied up Allied forces, as Dönitz was keenly aware,[180] drastic retrenchment would not have been a good idea.

Morse and Kimball suspect that only a "criminal lack of liaison between the German air and naval technical staffs"[181] explains the 6-month delay between the Luftwaffe's acquisition of an intact S-band radar and the

Figure 47
1944 Effects of Retrenchment

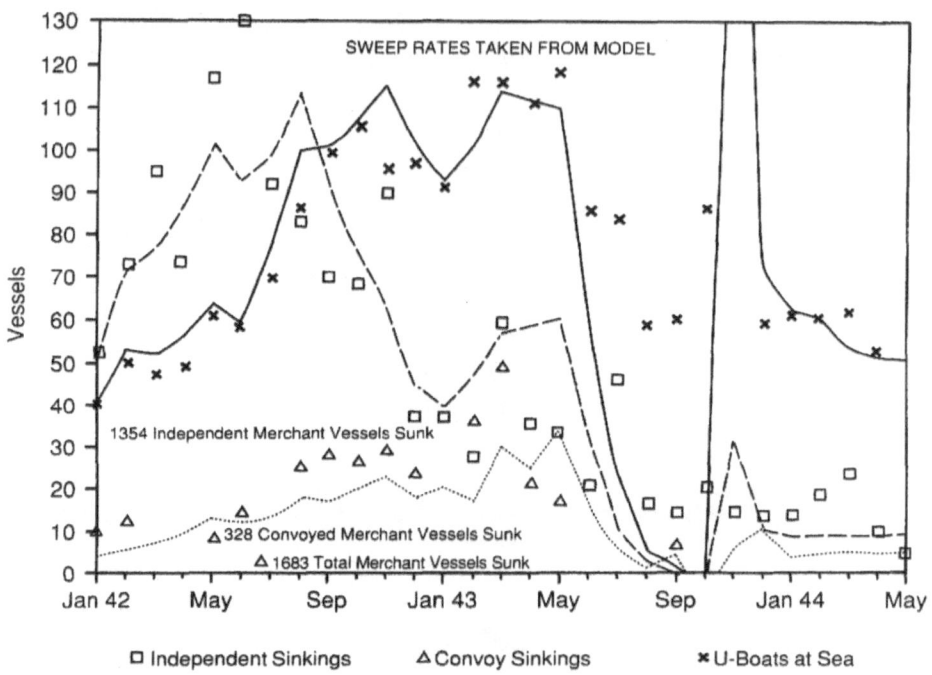

● *Line Equals Calculated Values; Symbols Equal Actual Values.*

U-boat command's realization that this radar accounted for the heavy U-boat losses aircraft inflicted. The S-band "Rotterdam Gerät," discovered in March 1943, remained unknown to the Kriegsmarine until September, so the first Naxos receivers did not go into use until October. What if communications had been perfect? Naxos could have gone into use in April 1943, but as the U-boat circulation model shows, only seven more merchant vessels would have been sunk as a result. (See figure 48.)

The Germans might have simply adopted the maximum submergence policy permanently: this policy worked well against the ASV Mark II—better, in fact, than the Metox countermeasure for which it was a stopgap—and it worked well again against the ASV Mark III. What if Dönitz had introduced it earlier, or had kept it from "wearing off" after he first ordered it in 1942?

Figure 48
Early Naxos Deployment

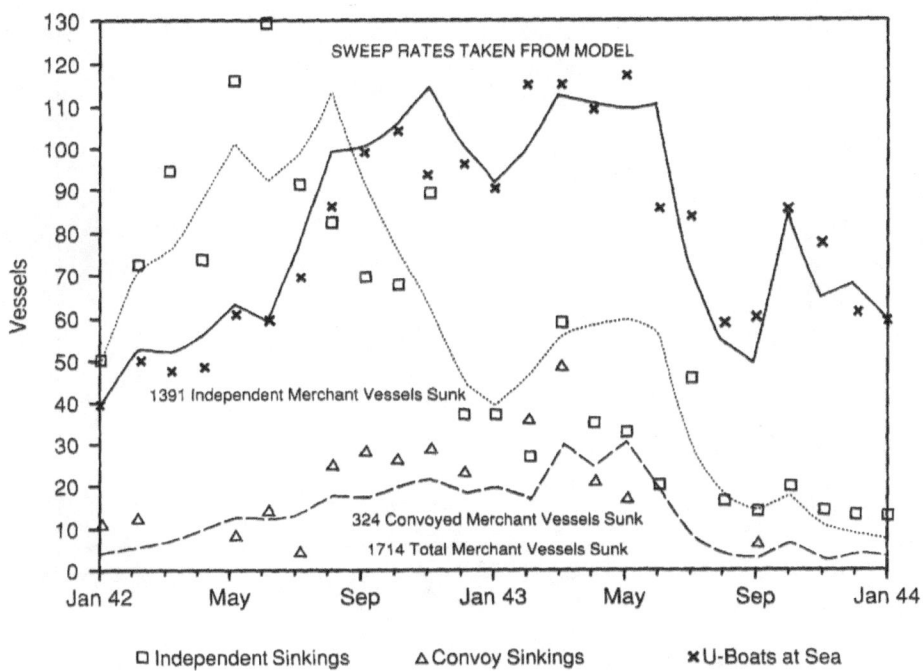

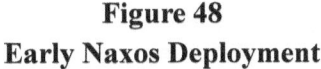

● *Line Equals Calculated Values; Symbols Equal Actual Values.*

The disincentive for maximum submergence is greatly increased transit time, with a concomitant decrease in time on station. The omnibus model shows that the increased survivability does not make up for the loss of time on station: almost 40 fewer merchant vessels would have been sunk by such a maximum-submergence campaign than were actually sunk. (See figure 49 and table 22.)

Going to the other extreme, Dönitz, reacting to the increasing backlog of U-boats awaiting repair, might have decided he could afford to lose a few boats if that was the price for getting them to sea promptly. He could have continued the medium-submergence tactics in use when Leigh Light flying began, sinking 85 extra merchant vessels. Note that this, the simplest of policy decisions—staying with the status quo—would have sunk a number of merchant vessels comparable to the bonus resulting

Figure 49
Maximum-Submergence Campaign

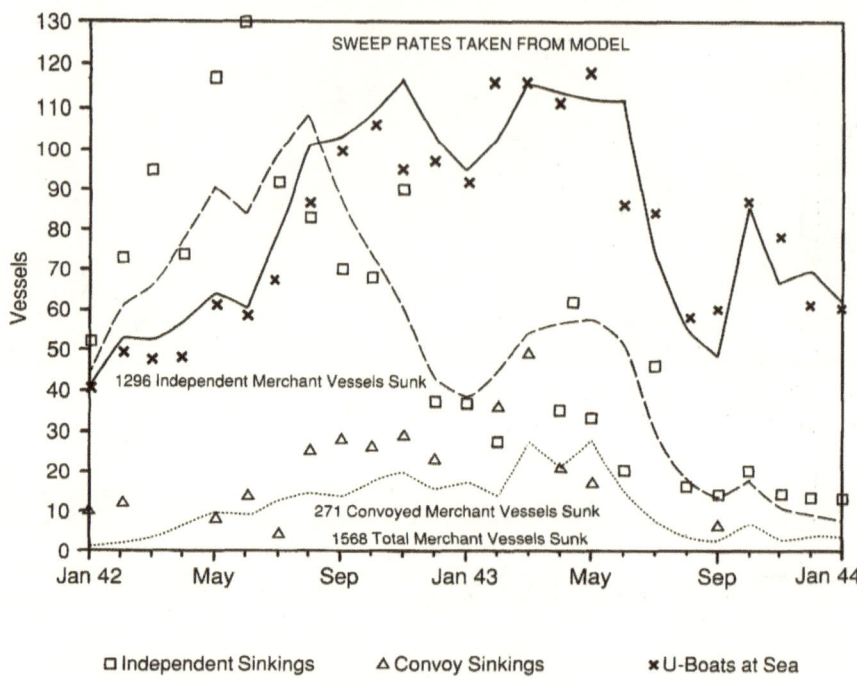

□ Independent Sinkings   △ Convoy Sinkings   ✗ U-Boats at Sea

● *Line Equals Calculated Values; Symbols Equal Actual Values.*

Table 22.—Results of Excursions Using the Omnibus Circulation Model

Case	Merchant Vessels Sunk	
	total	*vice base*
Actual History	1650	− 57
Maximum-Submergence Campaign	1568	− 39
Model Base Case	1707	0
Early Naxos Introduction	1714	+ 7
Medium-Submergence Campaign	1792	+ 85
Surfaced-Passage Campaign	1822	+115
Retrenchment Results through May 1944	1682	− 74
Base Case Extension to May 1944	1756	0

Figure 50
Medium-Submergence Campaign

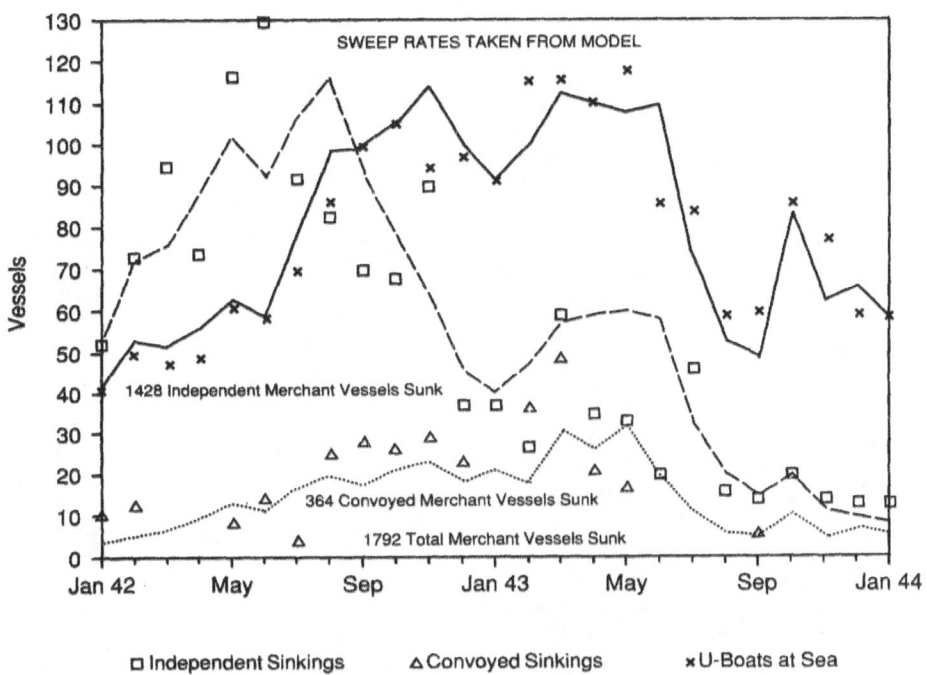

□ Independent Sinkings   △ Convoyed Sinkings   × U-Boats at Sea

● *Line Equals Calculated Values; Symbols Equal Actual Values.*

from fighting the whole campaign in the absence of convoy escorts, had that been possible. (See figure 50.)

Dönitz could even have instituted a radical policy of surfaced Bay transit. Doing so would have sunk 115 extra merchant vessels, all other things remaining equal. One might well ask if Dönitz would have been about to run out of submarines at the end of this scenario. He would not; in fact, he still would have had an increasing maintenance backlog piling up in France until D-Day. (See figure 51.)

Figure 51
Surfaced-Passage Campaign

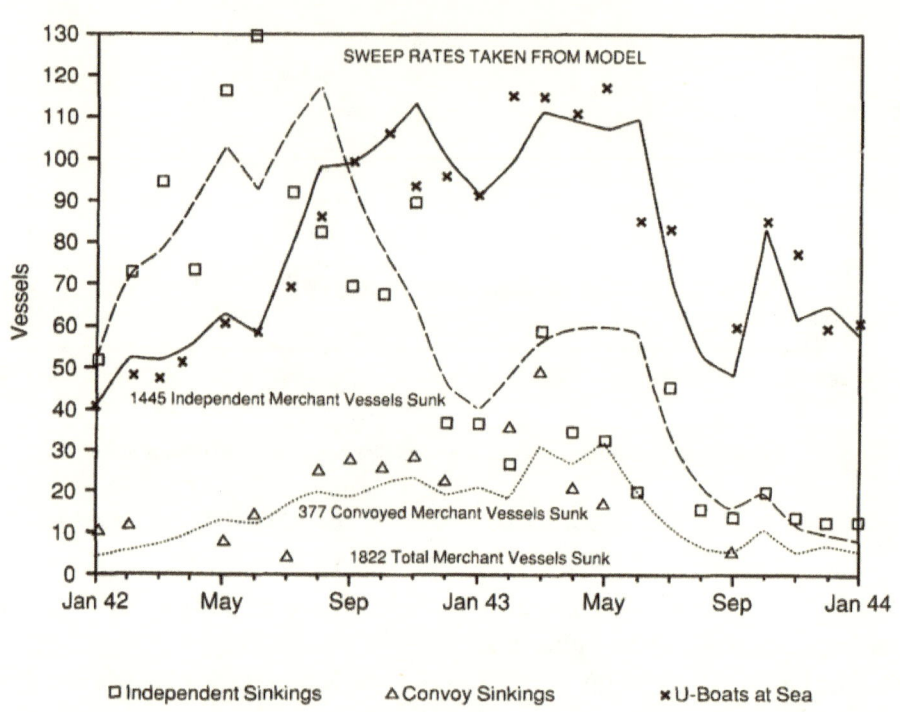

● *Line Equals Calculated Values; Symbols Equal Actual Values.*

8

Historical Conclusions

There are many who will be inclined to cavil at any mathematical or semi-mathematical treatment of the present subject, on the ground that with so many unknown factors, such as the morale or leadership of the men, the unaccounted merits or demerits of the weapons, and the still more unknown "chances of war," it is ridiculous to pretend to calculate anything. The answer to this is simple: the direct numerical comparison of the forces engaging in conflict or available in the event of war is almost universal. It is a factor always carefully reckoned with by the various military authorities; it is discussed ad nauseam in the Press. Yet such direct counting of forces is in itself a tacit acceptance of the applicability of mathematical principles, but confined to a special case. To accept without reserve the mere "counting of the pieces" as of value, and to deny the more extended application of mathematical theory, is as illogical and unintelligent as to accept broadly and indiscriminately the balance and the weighing-machine as instruments of precision, but to decline to permit in the latter case any allowance for the known inequality of leverage.

—F.W. Lanchester, 1916[182]

    This chapter presents conclusions about the U-boat campaign itself, based on the statistical analyses of the sweep rate and on the Bay search model and the U-boat circulation model. Methodological points appear in the next chapter, and broader conclusions about countermeasures and measures of effectiveness in the last.

Two Unlikely Implements of Bay Search

Regardless of the search method used, the Allies struck at the basic weakness of the U-boat: it could not remain submerged indefinitely. As more than one source points out, it was this fundamental fact that enabled airplanes to prey upon U-boats at all.[183] To the extent that the submarines could increase the percentage of the Bay transit made underwater, they could make themselves immune to air search.

U-boats could not, however, cross the Bay with impunity because they could not cross it completely submerged. The Bay transit model enables us to find how safely U-boats could cross the Bay during various phases of the war, given the equipment prevailing on both sides at those times.

Nor could the U-boats maximize submergence without diminishing their operational effectiveness. The U-boat circulation model illustrates how boats obeying maximum-submergence orders in crossing the Bay spend so much time in doing so that they require at-sea refueling to be effective in their mission against merchant shipping. A compensating factor, not taken into account, is that the slower passage *may* have consumed less fuel,* so that boats that crossed under maximum submergence may have been able to stay at sea longer. However, this effect, if present, must have been slight: later in the war boats operated from Norwegian ports, passing to the Atlantic on voyages similar to those of maximum submergence in the Bay, and they had about the same endurance as we have imputed to the Biscay boats.[184]

At-sea refueling confers several advantages. As mentioned above, it saves time that boats would otherwise have to spend "commuting" to and from the patrol area. Refueling is the ultimate countermeasure to

---

* The slower passage consumes electric power more economically. On the other hand, the effect is lessened because the submerged use of electric power is inherently far less economical than the surfaced use of diesel: not only were the boats' hulls optimized for surface operation, but also the extra steps of converting diesel fuel into propulsive power via electricity (turning a generator and later, after battery storage, turning a motor) entail a great loss of efficiency compared with direct diesel propulsion.

offensive search in the Bay, because each at-sea refueling eliminates two Bay transits per U-boat refueled: 20 transits per U-boat tanker cruise. Moreover, keeping the boats at sea keeps them out of the repair process: as the U-boat circulation model shows, boats that return to France can spend many months waiting for repairs before venturing out into the Atlantic again.

The defeat of the tanker U-boats in fact left Dönitz with little choice but to withdraw his submarine force from the North Atlantic in the late summer of 1943. Indeed, the U-boat circulation model shows that a considerable apparent retrenchment follows automatically from the loss of the tanker U-boats.

The Allies used Enigma decryption to sink tanker U-boats. Without the tankers, combat U-boats had to run the risk of increased surfacing in the Bay if they were to cross quickly enough to have time left to pursue their mission in the North Atlantic. Therefore we may say that while the most important German aid to Bay transit was the tanker U-boat, because it obviated some transits altogether, the most important Allied aid to Bay search was the Enigma decrypt, because it enabled destruction of the tankers and forced the U-boats to make more of their Bay transit on the surface.

The U-boat circulation model permits encounters with independent shipping anywhere except in the Bay itself, whereas convoys can be hunted only in the mid-Atlantic, requiring additional transit time. Encounters with independent shipping, however, declined over the course of the war, so that by the time the U-boats lost their tankers there were almost no independent merchant ships to sink during the scant time available at sea. The U-boat circulation model portrays these effects well. However, the post facto nature of the model's treatment of independent shipping—estimating coefficients by regression only—prevents investigation of the degree to which convoying complemented Enigma decryption by forcing the U-boats to concentrate on the mid-Atlantic region of safe convoy-hunting even while their at-sea time was cut so drastically by the loss of their tanker submarines.

The Dominance of the Mundane

As we have seen, the degree of submergence and the time at sea affect the outcome of the campaign much more than do the electronic countermeasures. Allied flying in the Bay was effective primarily because it kept the U-boats submerged and cost them time at sea. Tankers extended that time, and eliminated some Bay transits altogether.

Many of our observations in this study hinge on the fact that the huge maintenance backlog—of which Dönitz was painfully aware,[185] repeatedly petitioning Hitler to make more workers available—mooted any benefits of surviving the return trip. Recall that Morse and Kimball's original U-boat circulation model treated everything but repairs in a perfunctory way, and treated repairs with the surprising—yet accurate—"notions box" model. In their treatment of port congestion, the wartime operations researchers seem to have grasped an issue neglected by postwar writers, who have mathematically addressed almost every other aspect of queuing—a major area of operations research.[186]

9

Procedural Implications

> *Certainly the analysis of the confusing complexities of actual operations into such components is essential; but this dissection must not just hack through organs: it must separate them, and having become clear concerning the individual parts, must understand their synthetic recombination as an organism—an entity having a structure of its own, going far beyond the sum of its parts.*
>
> —Bernard Osgood Koopman, 1980[187]

Many of the results of the preceding analyses of Bay search and U-boat circulation are really statements about procedures in military modeling and analysis, rather than about Second World War antisubmarine warfare. Because the war is over, these lessons—seemingly dismissable as "mere methodological points"—may be more important than anything revealed about how to fight U-boats.

Simple Models Can Work!

Perhaps most significant in today's environment of massive computer models is the fact that the simple models used in this study replicate historical events quite well, and do so without relying on large numbers of *post hoc* curve-fitting parameters. (Recall that the regression-determined independent merchant-shipping losses are an optional extra in the U-boat circulation model.) The degree of independence from after-the-fact knowledge suggests that the many similar models* in today's literature could aspire to some accuracy in predicting the results of similar future conflicts. The most obvious example of such a conflict is another resupply of Europe during a conventional war: NATO forces convoy merchant shipping across the North Atlantic while Soviet submarines operating from bases on the Kola Peninsula try to sink them. Passage of the submarines through the Greenland-Iceland-United Kingdom (GIUK) gap parallels U-boat passage through the Bay of Biscay, and the battle is one of attrition, carried out against the submarines by "barriers" at several stages in their circulation. Another example is the operation of Soviet submarines from Vladivostok, whose choices for escape from the Sea of Japan would be limited.

The desire for simple quantitative models of military operations is not restricted to those who study the submarine threat to sea lanes of communication. ICBM exchange models[188] work from hardware characteristics to operational outcomes, as do the models presented here. The demonstrated ability to chart the course of a U-boat campaign suggests that modeling a strategic nuclear exchange is not an impossible task. Although one can argue that existing models have oversimplified certain issues (many, for example, ignore command and control considerations and damage mechanisms other than blast[189]), these issues seem amenable to quantitative treatment.

Simple models of conventional war on land present a different sort of problem altogether. Many more types of hardware and far more people interact, and terrain—in all its variety—plays a leading role in determining the outcome.

In the past 20 years, the formulation of simple models has come to include modeling one or more aspects of our entire planet. Such models—addressing, for example, global economic or ecological processes—bear mathematical similarity to the U-boat circulation model in that they compute state histories from discrete analogues of systems of

---

*   Such submarine circulation models appear, albeit with somewhat notional parameter values, in Nitze and Sullivan's *Securing the Seas*, Enthoven and Smith's *How Much is Enough?*, Bart and Cohan's *Model of Anti-Convoy Effectiveness*, and Mihori's *U.S.-Japan Security Policy*. The notional values could be replaced by values coming from ancillary models, analogous to the submodels of Bay search and convoy combat used to provide values for the U-boat circulation model presented here.

coupled differential equations. Critics have harped upon the nonverifiability (among other defects) of these global models, and some have attempted to tar all "systems-analysis"* efforts with the same brush.[190] Demonstrating verifiability—and in fact verity—in a formally similar model will, perhaps, act in at least some small measure to cleanse the systems-analysis escutcheon.

A philosopher of science would point out that the conformity of the models' results presented here to historical reality does not "prove them true."[191] The test against history could, however, have cast doubt upon them, and is thus a wicket through which such a philosopher would require that they pass. Some analysts shy away from such verification tests on the grounds that chance might make a historical outcome differ from a modeled one even though the model is not wrong,[192] or that the historical outcome might have resulted from a factor the model did not include.[193] These lines of reasoning do not persuade me that the historical test ought not to be made, only that the model ought to have the right to appeal an unfavorable outcome.

For models featuring regression-derived coefficients, verification is inextricably woven into the process of determining the values of the coefficients, as in the case of J.H. Engel's analysis of the battle of Iwo Jima in terms of the Lanchester "Square Law."[194] The modeled results fit history, but history provides time-series data only for the number of Americans in the fight—the daily numbers of Japanese and the two sides' troop qualities are all results of curve-fitting. Having survived such a test, a regression model—or any model entailing "tuning"—runs the risk of being perceived as an *ad hoc* theory—defined as one such that "the facts supporting it are only those it was introduced to explain"[195]—a situation summarized by the apocryphal quote, "We have calibrated our model to the results of the 1973 Arab-Israeli War and we can reproduce its results."[196]

Finally, the models used here to analyze the U-boat conflict contain no nonoperational variables, or operational variables with values based

---

* A useful distinction divides "operations research" or "operations analysis" from "systems analysis," reserving the first two terms for the quantitative study of ongoing or past operations, and the last for the study of future or hypothetical systems. In such usage the systems analyst, by definition, works with fewer facts and thus has a harder job than his or her operational counterpart.

on "judgment." These sorts of variables, such as "attacker's equilibrium attrition rate"[197] and aggregate firepower,[198] arise in models of land warfare and raise many of the doubts surrounding these models.*

Formulation of Measures of Effectiveness Today

As Morse and Kimball point out, almost any kind of operations-research analysis requires a measure of effectiveness. They provide examples, including but not limited to those used in this study. The modern reader, however, will note many differences between the measures of effectiveness Morse and Kimball advanced and those appearing in today's literature.

The Second World War bombing, antisubmarine, and antiaircraft campaigns analyzed by the original operations researchers presented a limited range of possible measures of effectiveness. In the antisubmarine campaign, for example, the Allied side could choose among such measures of effectiveness as the number of U-boats sunk, the number of successful Atlantic merchant-ship transits, the number of merchant ships saved, the number of tons of cargo saved, or the number of tons of cargo delivered to Britain. Each of these measures would have led to a different approach to antisubmarine warfare. In fact, the First World War argument over whether to use surface vessels to escort convoys or to hunt U-boats can be seen as a dispute over measures of effectiveness: to save the maximum number of merchant vessels, the surface vessels should escort convoys, whereas to sink the maximum number of U-boats, the surface vessels should hunt U-boats.[199]

Today, in peacetime, we have difficulty creating satisfactory measures of effectiveness because we lack the specifics of the war in which the weapon will be used. For example, today's ASW weapons could be used in an Atlantic context much like that of the Second World War, but they might also be used to attack ballistic-missile-launching submarines.

Another difficulty is that many, perhaps in some sense all, of our weapon systems have deterrence as a primary mission.[200] Because "wars averted" is not a knowable quantity, the search for surrogates leads to a wide variety of measures of effectiveness. In today's peacetime military operations analysis, measures of effectiveness retain their recognized

---

* Stockfisch treats doubts about firepower scores comprehensively.

importance, but their development is seldom carried to completion. Many articles whose titles give the impression that they will tell how to measure something's effectiveness do not actually present any such measure. Instead, they propose criteria for selection of measures of effectiveness, or even develop frameworks selecting criteria for determining for measures of effectiveness.

The strategy of deterrence requires not only that one have the ability to inflict unacceptable damage upon an adversary in war, but also that the potential adversary recognize that capability and therefore decide not to start a war. Therefore, weapons and plans must be effective not only according to one's own measure of effectiveness, but also according to that of the potential adversary. Some writers stipulate that the adversary may act according to some unknown form of alien logic.[201] Though they thus avoid the trap of considering him to be a mirror image of themselves, they cannot ever satisfy themselves that any proposed measure of effectiveness would make sense to the adversary. Investigations of the other side's operations-research literature[202] can obviate that concern, except among another group of skeptics: those who see such literature as intentionally deceptive.

Reasonable people do not differ as much as some proponents of the "alien logic" theory would have one believe, and any differences they have are more likely to be of values than of logic.* Dönitz and the Allied operations researchers did not differ in logic or values, eventually concluding that success in the North Atlantic campaign was measurable in cargo ships sunk, regardless of their location or cargo, if any. One could argue further that if the two sides did not have *similar* measures of effectiveness, they would have nothing to fight about. In the period of

---

* An unreasonable person might follow an "alien logic," but with little profit. Adolf Hitler, not widely cited as a rational actor, arguably acted according to alien logic during the latter part of the war. The evidence for such an argument, however, would include no successes: his East Front no-retreat policy (Guderian, p. 256) and his insistence that the Messerschmitt-262 fighter be developed as a bomber and never even be referred to as a fighter (Galland, pp. 258-262) followed a logic of quixotic suboptimization alien even to his own field commanders, and doomed to fail in the real world. (Galland comments, "One might as well have given orders to call a horse a cow!")

the Second World War known as the "phony war," for example, France measured success in the war against Germany in terms of French land not conquered—as the construction and use of the Maginot line indicate—yet Germany did not measure success in terms of French land taken.[203] Thus the belligerents fought lackadaisically from September 1939 until May 1940, when Hitler turned to the conquest of France. Then Germany and France began to operate according to the same measure of effectiveness, French territory, and fighting began in earnest.

Second World War Measures of Effectiveness Did Not Beg the Question

The measures of effectiveness used by Second World War operations researchers, and carried through in the present work, did not beg any questions. Although one might quarrel with the use of merchant vessels sunk as a criterion, one could hardly argue that it is a vague, abstract, or subjective measure.

In today's practice, alleged measures of effectiveness spring up full-grown as inputs to models. For example, many strategic-exchange models that include active defenses assume that the defensive systems can be characterized by how many "kills" they produce. The trouble with this measure of effectiveness is that it sweeps a difficult part of the problem under the rug: 1,000 interceptors with a kill probability of 0.5 pose an allocation problem not presented by 500 perfect interceptors. Against 1,000 targets, to be sure, the 1,000 50-percent effective shots would produce, on average, the 500 kills guaranteed by the 500 perfect shots. Against 500 targets, however, the latter force would still produce 500 kills, while the former could hope to average only 375.

Even worse is the situation in which model designers start with an idea such as "target value,"[204] build the model around it, and then—in effect—sit back and say, "Now just give us the data." The key distinction between this sort of question-begging measure of effectiveness and a useful one is that the useful one can be measured.

Intellectual Get-Rich-Quick Schemes

Closely related to the mistaken view that one can simplify one's work by calling some chosen input a measure of effectiveness is the view that any of several methodologies, such as cluster analysis, the Delphi method,[205] or Saaty's analytic hierarchy process,[206] can create information out of thin air. These approaches are valid means of collecting information scattered amid large amounts of clutter, but they do not evade the "garbage in, garbage out" principle any more than a set of wind chimes evades the Second Law of Thermodynamics.

Composite Measures of Effectiveness

One often sees attempts to measure effectiveness as the sum or product of several desirable quantities, perhaps with undesirable quantities subtracted away or placed in the denominator. These formulations have the merit that an increase in a good quality, or a decrease in a bad one, will increase the measured effectiveness. (Not all formulations have even that trait. One occasionally sees cost comparisons of aircraft done in dollars per pound,[207] even though both cost and weight are undesirable.)

The formulator faces a choice between addition and multiplication. Multiplication allows combining unlike terms, such as firepower and mobility (assuming these can be defined). Multiplication also allows low-scoring qualities to drag down high-scoring ones: in particular, a zero score in any one factor leads to a zero overall. Addition lacks this property, a fact not always noticed. For example, an introductory book[208] on sonar presents the equation

$$\text{Tactical Sonar Performance} = \text{Sonar Equipment Performance} + \text{Ocean Transmission Performance}$$

which fails to take into account that if the equipment is of zero quality, performance will be zero no matter how favorable the ocean transmission.

In some cases, however, addition is appropriate because the ingredients are all of like units. The question of assigning weights to the addends then arises. Operations-research texts properly devote little attention to this problem, except perhaps to point it out as a difficult but nonmathematical precursor to such methods as linear programming.[209] The practical worker, on the other hand, soon finds that any method requiring a weighting scheme raises the question of what weights are to be used, and that all sorts of imponderables can be dispatched neatly if one is allowed to assign *a priori* weights to various quantities and then take their sum.

Even more difficulties evaporate if the weights are used to combine incommensurables.*

These points apply to the case in hand because this study does not use any arbitrary *a priori* weights. Also, only three *a posteriori* curve-fits appear, all sub-models.** Yet many discussions in today's defense consulting community seem predicated on the assumption that the guessing and finding of *a priori* and *a posteriori* weights or functional forms is the whole art of operations research. Salient examples occur in the literature of wound ballistics, in which one source gives the probability that an incident bullet or fragment of velocity *v* and mass *m* will be fatal as

$$P(Kill|Hit) = 1 - e^{-a(mv^{\beta} - b)^{n}}$$

where *m* is mass, *v* is velocity, and the rest of the variables on the right-hand side (and indeed, the functional form of the right-hand side) are chosen so as to fit data from experiments and/or battlefield experience.[210]

Habituation to such expressions can lead to loss of appreciation of analytical content where it is present. For example, the "countermilitary potential" of a strategic nuclear warhead is defined as yield$^{2/3}$/CEP^2,

---

*   Two articles (one by Pankin, the other by Cover and Kellers) on the measurement of baseball players' offensive effectiveness show the application of this method and a clever alternative. Pankin uses a variety of statistical and logical arguments to modify the traditional slugging average into a more complicated weighted combination of singles, doubles, triples, homers, at-bats, hit-by-pitches, and so on that correlates better with average runs per game than do the traditional batting and slugging averages. Cover and Keilers simply construe the events of a player's record (in order) as the record of a whole team, and accordingly construct a series of synthetic games played by this team, thereby finding "the number of earned runs per game [the player] would score if he batted in all nine positions in the line-up" by sheer score-keeping. The reader may decide which approach is more appealing: the latter correlated even better with average runs per game than did the former.

** The relationship of independent merchant ships sunk to U-boats at sea, the U-boat/merchant vessel exchange rate for convoyed merchant vessels, and the contribution of signals intelligence to German search for Allied convoys, as analyzed in OEG Study 533.

where yield is the energy of the weapon and CEP (circular error probable) is the median miss distance of the guidance system. So defined, the countermilitary potential fits into an equation, derivable from the physics of the situation,[211] expressing the probability that the warhead will destroy a small target of great sturdiness, such as a missile silo. Yet one sees allusions to the countermilitary potential that present it as a (perhaps well-chosen) measure,[212] index,[213] or "indicator"[214] without any particular empirical underpinning.

"Laws" vs. "Models"

As has presumably been noted by many of their latter-day readers,[215] early operations researchers presented their discoveries as "laws" and "theories," not "models." Today's operations researchers refer to almost everything as a "model" and would never dare declare any finding a "law," except in jest.* While such a change in nomenclature can be justified on the basis of today's philosophy of science,[216] it can make presenting results to nonspecialists more difficult.

For example, while the nonspecialist reader of a study might suspect a "Poisson law" of being an ivory-tower abstraction or a "Poisson distribution" of being unduly technical, reference to a "Poisson model"[217] invites immediate skepticism. The term, the methodological upheavals in physical science that created it, and the diffidence with which particular models are advanced[218] combine to give the impression that no model is any better than any other.[219]

The belief that all models are created equal has an adverse effect on the producers as well as the consumers of studies. One may all too easily create an impressive-looking study based upon impressionistically chosen weighting factors and functional forms. The consumer, in turn, detects the absence of any meaningful derivation and concludes—perhaps correctly—that the study is not useful because it only embodies the author's preconceptions. For example, a study recommending larger convoys on the basis of an objective function that was the weighted sum of such judgment-derived quantities as "defensibility" and "capacity" would not be likely to displace a decisionmaker's predisposition against "putting all one's eggs in one basket." By contrast, a study—based on 10th-grade

---

* As in Norman Augustine's delightful book, *Augustine's Laws*.

geometry—showing that the escort requirement depends on the square root of the number of merchant ships defended is compelling and not unduly abstruse.

Strangely, today's "models" more strongly resemble "laws" than do many of the "laws" of the wartime operations researchers—their various coefficients cannot be interpreted other than as constants of nature. For example, in the Poisson coverage formula, or "law of random search,"

$$\text{Probability of Sighting} = 1 - e^{\text{-search effort/area}}$$

"area" is the area of the region searched and "search effort" is the product of sweep width, speed, and hours flown. On the other hand, in the "model"[220] for the probability that a small-arms or fragment hit will be fatal,

$$P(\text{Kill}|\text{Hit}) = 1 - e^{-a(mv^{\beta} - b)^n}$$

m is mass and v is velocity, but whereas β admits of some physical interpretation, a, b, and n do not, and all are chosen purely empirically. Moreover, their *dimensions* are chosen empirically: if β is determined to equal 2, for example, then b—and the nth root of $1/a$—must also stand for energy for the equation to balance dimensionally. Wartime operations researchers did not usually work in such a fashion; though they did on occasion derive a numerical constant from experience or experiment, they did not create functional forms or dimension-laden conversion factors On an *ad hoc* basis.*

---

*   A seeming exception showcases the rule. Sternhell and Thorndike, in discussing attacks on submerged U-boats by surface ships using sonar, address sources of uncertainty as to the location of the submarine during the "blind time" t between the attacking ship's last sonar fix and the attack. These arise from errors in the fix, errors in the estimates of the submarine's speed v and course, and the submarine's ability to alter course and speed. The resulting expression (a polynomial in vt) contains constants of various dimensions (whose values the authors say can be derived "from physical characteristics of the gear"), but is based on kinematic reasoning, not impressionistic choice of functional form. (Sternhell and Thorndike, p. 117.)

If one did not mind redubbing so many existing "laws" as "models" and vice versa, one could establish a useful distinction between the two terms: a "law" contains one or more coefficients of empirically based dimension and no physical interpretation beyond the law's statement; a "model" contains only variables with well-defined pre-images in the real world. Thus the "law of gravity," expressing the attractive force F between masses M and m at range r as

$$F = \frac{GMm}{r^2}$$

qualifies as a law because the value, dimensions, and very existence of G arise only in this expression, whereas the inverse-cube visual-sighting "law"

$$\frac{\text{probability of sighting}}{\text{glimpse}} = \frac{kh}{r^3}$$

relating probability of sighting to airplane height h and ground range r ought really to be called a model, because k admits of an obvious interpretation, target cross-section.

Operational Sweep Rate and the Poisson Coverage Formula

The concept of operational sweep rate pervades early operations-research literature, but has strangely faded from prominence in today's literature and curricula. The author has often had occasion to invoke it when discussing more modern problems—such as attacks on mobile ICBM launchers—with professional audiences, only to find that they have never heard of it. This condition appears all the more peculiar when one realizes that a section of Morse and Kimball is reprinted in Newman's *The World of Mathematics* immediately adjacent to Lanchester's seminal paper introducing his "square law."[221] Since Newman is by far the most accessible source for the Lan-chester article, it is the most commonly referenced, and yet few people appear to have read the adjacent article on searching for submarines.

The Poisson coverage formula, or "law of random search,"

$$\text{Probability of sighting} = 1 - e^{-\text{search effort/area}}$$

has fared only slightly better, occasionally surfacing in articles about antisubmarine warfare[222] and being mentioned as an important fruit of the wartime labors, as well as occasionally being used *ad hoc* without any indication that it can be derived from anything.[223,224]

"Search as a General Operation in Practical Life"

Koopman uses the above phrase in pointing out the prevalence of search operations in human endeavor, giving examples such as searching for mineral deposits, missing persons, or criminals.[225] Without casting such a broad net, one may note the criticality of the search paradigm to an understanding of the U-boat war: today, with more accurate and deadly weapons, the operations of war are operations of search.

Moreover, some operations that would not normally be construed as searches can be treated with the same mathematics. The most obvious example is screening, as of a convoy by a curtain of destroyers or a city by anti-aircraft radars and guns. (Indeed, Koopman titled his book *Search and Screening*.) Minesweepers form a moving screen; the "penetrators," the mines, stay still. Barrages undertaken in ignorance of the targets' true positions can be viewed as *single-try* searches, in which the search plan is drawn up in advance and executed all at once.

Nor is submarine warfare the only venue in which search operations figure in more than one way. For example, much interest currently centers on piloted strategic penetrating bombers. These bombers* might well have to take off amid a barrage of submarine-launched ballistic missiles, fly through one or more air defense screens, and then search for their targets.

Analysis vs. Synthesis in C3I Measures of Effectiveness

The disappointment that follows from reading much of the work done in the defense-analysis industry in and around Washington, D.C., comes from the fact that these studies consist only of analysis, with no synthesis. The resulting "structure," perhaps presented as "not just an answer, but a

---

*   And their tankers, whose importance a student of the U-boat campaign would be unlikely to underestimate.

Figure 52
Typical Electronic-Warfare Diagram
Adapted from Fitts, p. 2

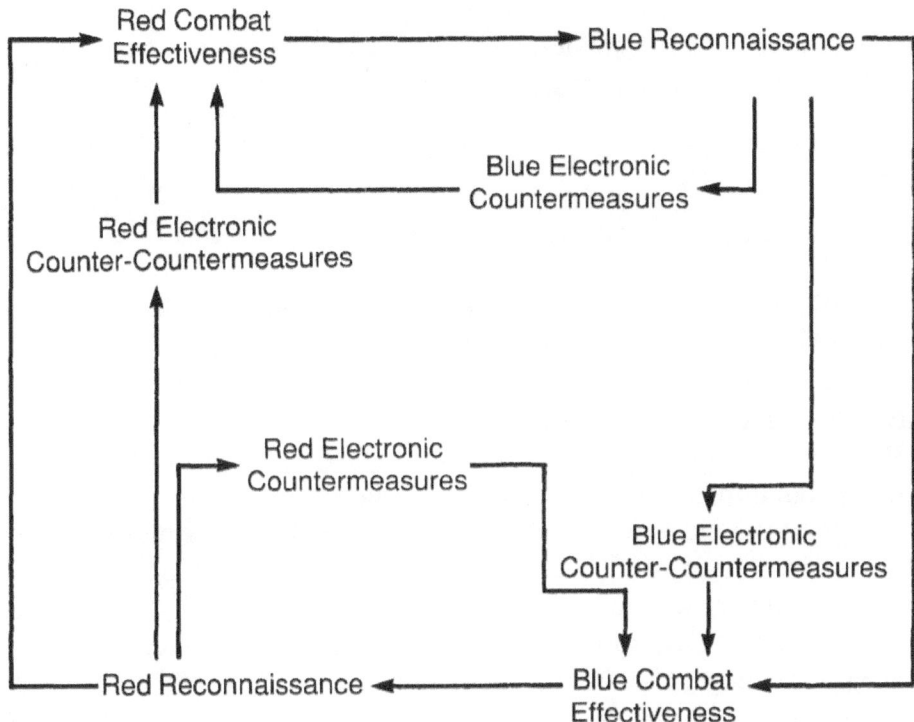

way to think about the problem," answers nothing and probably only codifies the obvious."*

Today's typical electronic-warfare article is accompanied by a flow chart in which almost all possible pairs of elements are connected. (See figure 52.) Without quantification to help the analyst decide what to

---

* For example, one such "approach," though said at the outset to integrate "the operational users' and analytic communities' efforts to scope and evaluate C2 problems with standard and evolving operations research tools," ends by saying, "Ultimately, [this approach] can be expected to identify measures sensitive to the architectural variables within C2 systems, which reflect mission effectiveness."

consider, what to emphasize, and what to ignore, present-day models often suggest that nearly everything is important and affects everything else.

The older sources cited in this study read differently: rather than sedulously dissecting the problem under study and then just leaving it "like a patient etherized upon a table,"[226] the authors of these older studies develop their structure, plant its lower extremities in a firm basis of fact, and work their way back up to an answer to the original question.

The Product of Quantitative Models

Many people see quantitative models as methods of calculating answers. In themselves, these answers rarely do much good. However, thoughtful consideration of a model's results can lead to what have been called "insights,"[227] "explainable surprises,"[228] or "novel facts."[229] For example, many have addressed the August 1943 retrenchment. The U-boat circulation model clarifies the point that a U-boat at sea in November cannot sink any more ships than it could have in August, so Dönitz's retrenchment did not gain him any merchant-vessel sinkings.

Lagrangian and Eulerian models (represented by Morse and Kimball's U-boat circulation model and the one developed here, respectively) portray events unfolding over time: the "state histories"[230] they produce are particularly likely to furnish insight. Other models, such as those of strategic exchange, capture the interaction of equipment parameters only in presenting a result. Such models, while a step above the notorious bean-counting methods, cannot give a sense of what is "happening" because they lack a time variable.

The Human Element

A final reason for the high quality of wartime operations research must be noted: the employment of geniuses. (The British Operational Research Section included two Nobel Prize winners and five Fellows of the Royal Society.[231]) The war effort plucked these individuals from their natural habitats and set them to work "seven days a week, 52 weeks a year."[232] Today's peacetime efforts—many of which are in fact window dressing[233] or rococo computer makework[234]—cannot hope to engage such talent with such intensity.

10

Generalizations

*[Our] volume on antisubmarine warfare [ASW] represents
a compromise between two major aims, to produce a unified
summary of the events and problems of the antisubmarine war
on the one hand, and to illustrate the scientific evaluation of
naval operations on the other. The approach is fundamentally
historical on both accounts, however, since the illustrations of
scientific evaluation are taken from various analyses and studies
made in connection with antisubmarine warfare during World
War II. Great care should therefore be exercised in making
predictions concerning the future of ASW from it. There is no
guarantee that the antisubmarine measures successful in the past
will continue to be adequate in the future. A clear understanding
of the events of World War II, their reasons and consequences, is
necessary, however, as background for any decisions which are
to be made in the postwar period. It is hoped that this volume
may serve to some extent as a convenient reference and source
of factual material. One overall conclusion is clearly evident
from it: the introduction of new weapons, gear, and tactics has
led to a continual interplay of measures and countermeasures
in which no other conclusion retains its validity for very long. If
this lesson alone is learned from it, the volume will have served
a useful purpose.*

—Sternhell and Thorndike, 1946[235]

One may generalize from the Second World War. For example, a future war in the North Atlantic, such as that Nitze and Sullivan envisioned, might include Anglo-American efforts to resupply NATO forces in Europe and Soviet efforts to interdict those forces. The Soviet submarine force based in the Kola peninsula would have to cross the Greenland-Iceland-United Kingdom (GIUK) gap in order to attack merchant shipping; one can surely draw parallels between such a campaign and the Second World War campaign studied here, with the GIUK gap in place of the Bay of Biscay. However, generalizations of an even broader nature can be made.

The Fallacy of the Second-to-Last Move

Discussing the postwar competition in strategic nuclear weapons, Herbert York identified "the fallacy of the last move,"[236] in which a proposed innovation is touted as capable of providing a permanent advantage. The fallacy, of course, lies in the fact that a strategic arms race is not a game with a last move: the opposition will not accept permanent disadvantage but will strive for, and eventually create, some improvement that will bring a return to the status quo or even a position of reversed advantage. Supporters of various ballistic missile defense proposals of the 1960's, 1970's, and 1980's have been accused of subscribing to this fallacy, as they have promised defense only against current ballistic missiles, not future ones designed to penetrate defenses.

Identifying the "fallacy of the last move" has had the unfortunate result of fostering a fallacy of the second-to-last-move, which sees only one move deeper into the game and holds that no innovation is worth making because it will eventually be countered. The Bay of Biscay offensive clearly illustrates that a sequence of temporary advantages can be as useful as a permanent advantage.

The Solitaire Fallacy

One also sees instances of what might be called the solitaire fallacy, in which the opposition is treated as a natural phenomenon. In a search-related instance of this fallacy, authorities disparaged the effectiveness of balloon-mounted radars for use against drug smugglers after a single balloon, the first of a proposed chain of six across the U.S.-Mexican border, failed to catch any drug-carrying airplanes. The drug runners had switched to ground vehicles and shipments by freight container. "There's a wonderful madness to all of this," said a Capitol Hill critic of the program. "Just as Customs is getting its air balloons, all the bad guys are moving into containers." An operator of the balloon thought that the smugglers *might* be routing their remaining flights around the balloon, which could be seen for miles.[237]

The critic ignored the possibility that the smugglers' switch away from aircraft had been forced on them by the balloon—or by anticipation of the rest of the balloons—and that they paid some kind of price (in time, money, or other coin) for departing from what had been their preferred means of transporting drugs. The operator seemingly lent insufficient credence to the possibility that smugglers might avoid the balloon, which covered only a portion of the border and was not always in the air. Neither man viewed the situation as a contest with a rational and determined adversary upon whom costs could be imposed, and neither saw the problem in terms of a drugs-not-arriving measure of effectiveness.

Troubles With the Top-Down Design Approach in a Countermeasures Environment

Today's defense analysts often emphasize a "top-down" approach, which is almost a codification of the solitaire fallacy. In its own way, this approach can also foster the same paralysis as that induced by the fallacy of the second-to-last move. Top-down design, originally a computer software concept, starts with the desired result—found by a "requirements analysis"—and works backward to the means of attaining that result. Such an approach was once seen as the only way to write large computer programs from scratch: beginners who mistakenly work from the bottom up generate unmanageable "spaghetti code," which others cannot read and even the writers often cannot debug.*

Lately, the top-down design philosophy has spread to other areas of endeavor. The initial stage of work on a new aerospace system, for example, is to ask, "What do we want it to do?" An answer might be "Find submarines." Having determined the system's primary function, one asks, "What does it have to do in order to do that?" This recursive process—aided by flow

---

*   More recently, an alternative approach based on creating and refining prototype programs has emerged.

diagrams and even computers to help keep track of the diagrams—eventually results in a complete description of all the functions the system entails. Then, and only then, can design work begin: earlier mentions of design ideas are decried as premature.

The top-down methodology precludes consideration of the state of the art: it asks what is wanted, not what is possible. While it is arguably a reasonable approach to designing software, the realities of hardware design may negate such idealism. The top-down method also cannot answer questions of sufficiency, which properly belong to the realm of requirements. For example, the method provides no answer to "How reliably must the system find the submarines: with what probability of missing a submarine (at a given range) and with what probability of false alarm when no submarine is present?" Emerging questions of sufficiency—those that arise several steps down from the top—are even harder to answer: given a three-man crew, how many windows does the aircraft need for visual spotting of surfaced submarines? The design team sees the question as one of requirements, while the requirements group sees the question as an implementation detail relating to crew size, about which they have no stated requirement.

The disadvantages of top-down hardware design are rarely pointed out;[238] the sanctity of top-downism in the software world has resulted in the idolatrous worship of its image in the hardware world. Yet the top-down technique works very poorly on weapon systems. Questions of sufficiency abound, as do other questions that designers think relate to requirements and requirers consider implementation details. Given a requirement for a 1,000-missile ICBM force, for example, who is to say how many launch control centers there should be or, in the case of mobile ICBMs, whether the launch control centers should also be mobile? The weapon designers will probably view such questions as policy matters, just like the proliferation and mobility of the ICBMs themselves. Policymakers will view launch control centers as an important technicality best left to the experts.

Unaided, the designers—who lack the software engineers' ability to create and test cheap prototypes, but who have to arrive at an answer if they are to continue with the design process—will try to find an answer through pure reasoning. In a perhaps ill-informed debate the like of which the author has witnessed on more than one project, the designers will start at the top and work *up*, asking what the weapon's mission is, what the mission's purpose is, and so on until they get to a lofty goal such as undisturbed continuation of the American way of life, from which they will try to work back down to an answer to the question about launch control centers.[239] Some even explicitly reject the input of strategic thinkers at this stage.*

This top-up design process does not work very well. Computer models and simulations of the sort developed in this study (though perhaps not as simple) can help facilitate the upward movement from engineering data to performance, but not from performance to worth. Additionally, the application of top-down design to weapons raises issues not found in nonweapon applications. Weapons that must contend with countermeasures face—along with all their other requirements—the protean requirement to defeat enemy countermeasures. One's impression of these can change rapidly and is in any case shaped by one's worst fears. Interminable redesign or defeat at the hands of a "threat-of-the-month club," which deluges the design process with numerous notional enemy countermeasures, can result.

A weapon system whose only mission is to defeat enemy weapons—such as a search radar or a radar warning receiver—faces nothing but protean requirements, and therefore top-down design methods can lead only to frustration.

The top-down approach applied in its pure form to a sensor system would start with the question, "How can I unfailingly detect enemy platforms without ever suffering from a false alarm?" While a pragmatist might immediately propose a signature for which to watch and a sensor capable of doing the watching, a true practitioner of top-down design would reject such a response as technologically opportunistic (or worse, "solution-oriented"), and would hold out for a decomposition of the enemy platforms' phenomenologies, followed by a selection of which phenomenology to use and only then consideration of how a given signature might be detected.

In the foregoing scenario the pragmatic designer and the top-down designer would probably get to the same answer in different amounts of time. But countermeasures pose an even more severe problem for sensor systems than for weapon systems. One can always specify levels of signature reduction or jamming that a given sensor cannot counteract

---

* "Edward Luttwak should leave the arena of technology and weapon system requirements to the engineers and system developers."—Blair Stewart, Aviation Week and Space Technology, letter, September 19, 1988.

without an unacceptable increase in the false-alarm rate or decreasing the sweep rate. One response would be to improve the sensor technology. Another would be to proliferate sensor types, monitoring several phenomenologies and fusing the observations into a single assessment of target presence or absence. Neither method, however, lends itself to the pure top-down approach, which has such difficulty in addressing questions of sufficiency.

Some model of combat operations whose output is a measure of effectiveness—such as the U-boat circulation model developed here—can cut the Gordian knot of sufficiency. That model can also reveal the contest's two-sided, reactive nature, avoiding the fallacy of the last move and the solitaire fallacy.

There remains the matter of choosing an appropriate measure of effectiveness. Although we have analyzed the U-boat war in terms of merchant vessels sunk or saved, one could argue that the merchant vessels were an input to the Allied war effort and that an output, such as the date on which Germany would finally surrender, should be used instead. Such a measure of effectiveness would, of course, entail a far greater modeling effort: land and air combat models such as those of Joshua Epstein[240] would have to be used to simulate the Allied invasion of Europe.

Strategy and Tactics

According to an early distinction between strategy and tactics, strategy comprised the set of decisions made and acted on out of sight of the enemy; tactics comprised the rest.[241] The advent of modern weapons, such as carrier-based aviation, has permitted outright battles to occur at greater-than-visual ranges, and the modern version of the distinction holds that a strategic actor is one who does not have "contact" with the enemy.[242] "Strategic warfare" has corne to refer to war made against population or industry, and—latterly—to such war waged with nuclear weapons.[243] The term "operational art"—introduced by Soviet thinkers but well on its way to adoption by the West[244]—identifies a more recently highlighted layer, between strategy and tactics. A further useful distinction divides strategy and "grand strategy," the latter being the total scheme of national military endeavor, including economic elements.

A submarine antishipping campaign such as that prosecuted by the Germans in the Second World War occupies an ambiguous position amid these distinctions. Because it involves a strike against industry—merchant-marine shipping—it is strategic in the sense defined above. Yet, even as one can object that a "strategic nuclear first strike" is not truly strategic—on the grounds that it is merely a tactical allocation of weapons to targets[245]—one could surely argue that applying submarines to merchant-shipping interdiction is merely a tactical exercise as well.

I believe that the solution to this problem of categorization lies in the role of Grand Admiral Karl Dönitz and in the small number of links in the U-boat chain of command. Dönitz's style of operation put him outside the strict categories of tactician, strategist, or grand strategist. He was all these things simultaneously: he advocated the grand strategy of strangling Great Britain's economy, he chose the areas in which his wolf packs of U-boats would operate, mandated immediate changes in the methods they would use to evade detection in the Bay of Biscay, and often he directed their offensive movements on a real-time basis. Perhaps the term "battle management," currently in use with respect to the set of decisions made while a space-based ballistic missile defense system engages enemy missiles, best describes Dönitz's role.

Nor was Dönitz the only admiral to operate in such a manner: Yamamoto, Nimitz, and Halsey similarly exerted significant control from ashore.[246] Twentieth-century weapons (be they U-boats or ICBMs) and communications technology permit the personal implementation of a strategy by its formulator, so that a U-boat commander (or ICBM targeteer) is at once a strategic and a tactical actor—strategic in the sense of bringing the war to civilians, tactical in the sense of guiding individual weapon platforms.

"Wiser Air"

When working his or her way up from data to "insight," "explainable surprise," or "novel fact," the analyst (or, more accurately, the synthesist) must take care not to overshoot the original question. Doing so results in the "top-up" analysis mentioned above, in which the study participants repeatedly outdo one another in discerning that the study approach just proposed "doesn't tell us what we really want to know." At each stage, the previous apex is downgraded to being one of several important steppingstones one level lower than the ultimate goal. Because the study's resources do not increase, effort that could be spent answering the original question is frittered away. One manager contended with this problem by deriding questions of needlessly high level as asking, "Why is there air?"[247] In fact, he did so often enough that he came to contract the phrase to "wiser air."

Yet the process he so frequently bemoaned followed inevitably from an attempt to use top-down methods in the absence of a previously defined top. Such a true top, be it the intention to land a man on the moon and return him safely (as in the heyday of NASA) or the unconditional surrender of Germany and Japan (as in the Second World War) must—by its very nature—come from outside.

The Parallel Hierarchies

The hierarchy of military thinking (tactics, operational art, strategy, grand strategy) is a hierarchy of efforts to be effective, and thus is paralleled by a hierarchy of measures of effectiveness. (See table 23.) *One must be careful to select a measure of effectiveness appropriate to the level of operation.* For example, a simplified version of the U-boat circulation model presented here was used in a final examination. After being led to construct the model, the students were to select a measure of U-boat effectiveness, then use the model to evaluate various possible German countermeasures to offensive search in the Bay of Biscay. They had to justify their choice, but were advised that any reasonable measure would do if they could justify it well. Most, as expected, chose "merchant ships sunk per U-boat sunk" or "merchant ships sunk per month." One student, however, scoured the readings for the course until he found a statement in OEG Study 533[248] selecting operational sweep rate (of submarines for merchant ships) as a measure of effectiveness. Needless to say, his attempts to choose among Bay countermeasures on the basis of U-boat search effectiveness met with failure: he was trying to answer a high-level question with a low-level measure of effectiveness.

Table 23.—The Parallel Hierarchies

Context	World War II Example	Measure of Effectiveness
Grand Strategy	Unconditional Surrender	Time Until VE Day
Strategy	Establish Second Front	Time Until D-Day
Operation	Ship Goods to Britain	Tons Landed
Tactic	Defend Convoys	Exchange Rate
Tactic	Patrol Bay of Biscay	Sweep Rate $\cdot$ P(K|S)

We have seen how a critic of a certain course of action, such as offensive search in the Bay of Biscay or the use of balloon-borne radars against drug smugglers, can be blinded to the action's overall benefits by its apparent ineffectiveness at a low level. Similarly, an advocate of a certain course of action may make the claim that his improvement, being quite effective in terms of a lower-level measure of effectiveness, must be at least somewhat effective in the larger sense on the ground that "every little bit helps."

Both lines of argument suffer from the fallacy of composition, in which an attribute is imputed to the whole on the basis of its presence in the parts. Consider the example of maximum submergence: although it was a highly effective means of lessening the danger to U-boats in the Bay, it was counterproductive in terms of the ability of the U-boat force to sink shipping.

The effectiveness of efforts at deterrence is particularly difficult to measure because the obvious measure, wars averted, is unknowable. Belief in an escalation ladder has led to the belief that most major systems play a deterrent role, which may explain latter-day difficulties in arriving at reasonable measures of effectiveness for military systems.

Notes

1. Morse and Kimball, p. 3. Emphasis added.
2. W.S. Churchill, *The Second World War*, Vol. II, p. 598.
3. Morse and Kimball, p. 1.
4. J.A. Stockfisch, *Models, Data, and War: A Critique of the Study of Conventional Forces* (Santa Monica: Rand Corporation, 1975), p. 4.
5. Morse, "ORSA Twenty-Five Years Later," *Operations Research*, vol. 25, no. 2, 1977), p. 188. (As if any argument from authority were necessary for such an assertion.)
6. For example, see Hughes, *Military Modeling*, p. 285.
7. Thibault, p. 339, question 12.
8. Dönitz, *Memoirs*, pp. 193 and 195.
9. Quoted in Brodie, *A Guide to Naval Strategy*, p. 146.
10. Dönitz, *Memoirs*, p. 286.
11. Price, p. 97.
12. SRH-008, p. 209.
13. Morse and Kimball, Sternhell and Thorndike.
14. Dönitz, *Memoirs* passim, especially 2-4 and 235-6.
15. Blaug, p. 240; Blair, p. 7.
16. Hinsley, vol. II, p. 550.
17. A complaint voiced, for example, by Paul K. Davis on p. 180 of Din's *Arms and Artificial Intelligence*.
18. Dönitz, *War Diary*, April 15, 1942.
19. Sternhell and Thorndike, p. 90.
20. Dönitz, *Memoirs*, p. 110.
21. This section drawn from Sternhell and Thorndike.
22. Waddington, p. 15.
23. This section drawn from Price, pp. 59 and ff.
24. This section drawn from Price, p. 60 and following.

25. Sternhell and Thorndike, p. 30. Range information inferred from Price, especially p. 98.

26. Sternhell and Thorndike, p. 30.

27. This section drawn from Morse and Kimball, p. 95 & ff.

28. This account of early Metox taken from Gordon, p. 32.

29. Price, p. 95.

30. Price, p. 95.

31. Sternhell and Thorndike, p. 155.

32. This section from Morse and Kimball, p. 96.

33. Sternhell and Thorndike, p. 40.

34. Sternhell and Thorndike, p. 40.

35. This section from Sternhell and Thorndike, p. 156 & ff.

36. Sternhell and Thorndike, pp. 40-41, and Price, p. 121.

37. Hinsley, vol. III, part 1, p. 215.

38. This section drawn from Price, pp. 157-158.

39. This section drawn from Price, p. 189, and Sternhell and Thorn-dike, p. 153.

40. This section drawn from Price, p. 168.

41. This section drawn from Sternhell and Thorndike, pp. 57 & 156, and from Hinsley, vol. III, part 1, pp. 515-516.

42. Gordon, p. 31.

43. Price, pp. 169-170.

44. Price, p. 169.

45. Price, p. 187 and following, attributes the German S-band revelation entirely to the November crash. In this he is contradicted by Morse and Kimball, pp. 44-45 and 96, and Sternhell and Thorndike, pp. 155-158. All mention the voyage of Dr. Greven.

46. This section drawn from Sternhell and Thorndike, p. 157.

47. Dr. J.J.G. McCue, personal communication.

48. Sternhell and Thorndike, p. 158.

49. This section drawn from Sternhell and Thorndike, pp. 158-9.

50. Sternhell and Thorndike, p. 159.

51. Rossler, p. 196.

52. S-band range, Sternhell and Thorndike, p. 157; X-band range, p.159.

53. Hinsley, vol. III, part 1, p. 212.

54. This section drawn from Price, pp. 112 and 183.

55. Hinsley, vol. III, part 1, p. 217.

56. Hinsley, vol. III, part 1, p. 212, and Price, p. 71.

57. Hinsley, vol. III, part 1, pp. 212-213.

58. Hinsley, vol. III, part 1, p. 213, and Beesly as cited there.

59. Hinsley, vol. III, part 1, p. 217.

60. OEG Study 533.

61. Gordon, p. 57.

62. Waddington, p. 207.

63. Flying hours and Allied data from Morse and Kimball, p. 44, and Sternhell and Thorndike, p. 144. U-boats destroyed from Roskill. Dönitz data from *War Diary*.

64. Sternhell and Thorndike, p. 144.

65. Morse and Kimball, p. 44.

66. Sternhell and Thorndike, p. 144.

67. Sternhell and Thorndike, p. 142.

68. Hinsley, vol. II, p. 179.

69. Tidman, p. 59.

70. Hinsley, vol. III, pp. 224 and 217.

71. Sternhell and Thorndike, pp. 145 and 155.

72. Bay weather statistics from the *U.S. Navy Marine Climatic Atlas of the World,* passim.

73. As Ivanoff and Murphy discern in their paper "A Methodology for Technological Threat Projections of Soviet Naval Antiship and Surface to Air Missile Systems" in *Naval Power in Soviet Policy*, Paul J. Murphy, editor.

74. Hughes, *Fleet Tactics*, pp. 185-187.

75. See Modelski and Thompson.

76. Autocorrelation and the significance of autocorrelation terms is addressed in Bowerman and O'Connell, pp. 339 and following.

77. Bowerman and O'Connell, p. 343, and Knuth, vol. II, p. 64.

78. Guerlac, p. 714, text and figure. Also Keegan, p. 240.

79. A view mentioned by Stephen Meyer, p. 259 of Valenta and Potter.

80. P.M.S. Blackett, as quoted in Price, p. 67, or Morse and Kimball.

81. From vol. II and part 1 of vol. III of Roskill's *War at Sea.*

82. Waddington, p. 197.

83. Brodie, *Strategy in the Missile* Age, p. 126.

84. Waddington, p. 76.

85. United States Strategic Bombing Survey, *German Submarine Industry Report*, p. 17.

86. Morse and Kimball, p. 49.

87. Jones, pp. 409-410.

88. Morse and Kimball, pp. 49-50.

89. Sternhell and Thorndike, p. 84.

90. *Ibid.*

91. U-boat sinkings from Roskill; merchant vessels from Morison.
92. Washburn, p. 2-15.
93. Morse and Kimball, p. 43.
94. Koopman, 1946, p. 60.
95. Koopman, 1946, p. 57.
96. Waddington, p. 126.
97. Sternhell and Thorndike, pp. 140-141.
98. Sternhell and Thorndike, p. 159.
99. Gordon, p. 32.
100. Sternhell and Thorndike, p. 157.
101. Koopman, 1980, Chapter 3.
102. Koopman, 1980, p. 59.
103. As presented by, for example, Koopman (1980) in Appendix F.
104. Berkowitz, p. 16.
105. Brookner, pp. 82-85.
106. Koopman, 1980, p. 59.
107. Waddington, p. 124.
108. Koopman, 1980, p. 66.
109. Koopman, 1980, p. 66.
110. Koopman, 1980, p. 32.
111. Sternhell and Thorndike, p. 59.
112. Sternhell and Thorndike, p. 144.
113. Interestingly, Abraham Kaplan tells this tale as a methodological parable for social scientists in his *The Conduct of Inquiry*, p. 11.
114. Sternhell and Thorndike attribute the term to Operations Research Section—Coastal Command Report No. 204, *Air Offensive Against U-Boats in Transit*, December 10, 1942. (Sternhell and Thorndike, p. 145.)
115. Dönitz, *War Diary*, January 8, 1944.
116. Waddington, p. 233.
117. Waddington, p. 233.
118. Waddington, p. 234.
119. Terraine, p. 583.
120. Koopman presents, starting on p. 181, a much longer derivation without the simplifying assumption of non-diminishing returns.
121. Sternhell and Thorndike, p. 155.
122. From *Marine Climatic Atlas*.
123. Derivation explained in text.
124. Koopman, 1980, p. 71.
125. Washburn, pp. 2-6 to 2-7. Exclamation in original.

126. See any probability textbook, for example Ullmann, p. 114.
127. Sternhell and Thorndike, p. 155.
128. Price, p. 157.
129. Dönitz, *War Diary*.
130. Sternhell and Thorndike, p. 158.
131. Sternhell and Thorndike, p. 157.
132. Waddington, p. 238.
133. Dönitz, *War Diary*, entry for March 5, 1943.
134. Waddington, pp. 236-238.
135. Dönitz, *War Diary*, October 1943-January 1944, passim.
136. Dönitz, *Memoirs*, p. 230.
137. Morse and Kimball, pp. 52-53.
138. Terraine, p. 583.
139. This section drawn from Morse and Kimball, pp. 78-80.
140. Morse and Kimball, p. 78.
141. I am indebted to Warren Dew of NAVSEA for his help with this interpretation of the repair equation.
142. See, for example, Morse and Kimball, pp. 111-112.
143. Morse and Kimball, p. 80
144. Morse and Kimball, pp. 49-50.
145. National Security Agency SRM-008, p. 108.
146. See Nitze and Sullivan for a modern example.
147. Dönitz, *War Diary*, passim.
148. Rossler, page 166.
149. Sternhell and Thorndike, p. 144.
150. These figures were estimated from the fraction of boats at sea listed by Dönitz as being "on return passage." *War Diary*, 1942 and 1943, passim.
151. OEG Study 533.
152. Morse and Kimball, p. 47. N.B. conclusion that "the quantity k/n stays constant within the accuracy of the data."
153. Dönitz, *War Diary*, passim.
154. Dönitz, *War Diary*, passim.
155. Rossler, pp. 162 and 166-167. On page 162 Rossler refers to the tankers' ability to refuel a dozen Type VII C boats with four weeks' worth of fuel or five Type IX C boats with eight weeks' worth. These figures (and detailed table of all refueling operations on pages 166-167 of Rossler) combine to suggest that, in the context of a study such as mine, which does not attempt to differentiate among different types of U-boat, it is fair to say that a tanker U-boat could provide 10 "tankees" with an extra month's worth of fuel apiece.

156. Dönitz, *War Diary*, passim.
157. Roskill, passim.
158. Sternhell and Thorndike, p. 94.
159. Sternhell and Thorndike, p. 94. Koopman (1980) provides a thorough treatment of dynamic enhancement.
160. OEG Study 533.
161. See, for example, Pierce or Raisbeck.
162. Sternhell and Thorndike, p. 109.
163. Morse and Kimball, p. 47; Sternhell and Thorndike are probably working from the same data on their p. 106.
164. Sternhell and Thorndike, p. 108.
165. Sternhell and Thorndike, p. 107.
166. U-boats at sea are tabulated in the Dönitz *War Diary*. Roskill and Morison tabulate shipping losses.
167. See Gelb, edited volume, p. 84.
168. Gordon, p. 88.
169. Roskill, vol. II, appendix O, passim.
170. Fischer, p. 179.
171. Fischer, p. 16.
172. Brodie, *A Guide to Naval Strategy*, p. 144n.
173. York, "The Debate Over the Hydrogen Bomb."
174. Roskill, vol. II, p. 100.
175. Stern, p. 44.
176. United States Strategic Bombing Survey, *German Submarine Industry Report*, p. 26.
177. Such as Morse and Kimball, p. 45, and Hinsley, volume III, part 1, p. 217.
178. Dönitz, *War Diary*, August 1943.
179. Koopman, 1980, p. 22.
180. Dönitz, *Memoirs*, pp. 406-407.
181. Morse and Kimball, p. 96.
182. Lanchester, pp. 46-47.
183. Price, p. 182; Waddington, p. 31.
184. Dönitz, *Memoirs*, pp. 423-425.
185. Dönitz, *Memoirs*, pp. 166 and 230.
186. As in, for example, Saaty's *Elements of Queueing Theory*. "Random selection for service" is treated on pp. 243 and following, but in a very different style from that of Morse and Kimball.
187. Koopman, "Intuition in Mathematical Operations Research."

188. As presented in Bruce Blair's *Strategic Command and Control* and Davis and Schilling's *All You Ever Wanted to Know About MIRV and ICBM Calculations But Were Not Cleared To Ask*.

189. Blair, p. 39, mentions lack of consideration of command and control. His appendices A and B present "standard" damage calculations considering blast alone, whereas appendices C and D take electromagnetic pulse into consideration as well.

190. Saunders Mac Lane, letters to *Science*, p. 1144, vol. 241 (2 September 1988) and pp. 1623-4, vol. 242 (23 December 1988).

191. See Blaug, pp. 55-93.

192. Dalkey, in Quade and Boucher, pp. 251-252.

193. Hoeber, p. 145.

194. J.H. Engel, "A Verification of Lanchester's Law," *Operations Research*, vol. 2, May 1954.

195. Kaplan, p. 313.

196. Hoeber, p. 145, irony in original.

197. Epstein, *The Calculus of Conventional War*, p. 17.

198. Stockfisch, pp. 6-10.

199. Brodie, *War and Politics*, pp. 189-190.

200. Gray and Barlow, *International Security*, vol. 10, no. 2; p. 45.

201. Such as Alexander Kirafly, who writes in Earle's Makers of Modern Strategy, (p. 457) that "Japanese naval thought differs so radically from that of the western world that Japanese ideas of seapower cannot very well be stated in occidental terminology."

202. Such as that of the Soviet literature, conducted by Alan Rehm.

203. Kahn, pp. 334 and 408. The p. 334 passage is a quote from Gordon Waterfield, *What Happened to France*.

204. Found in the mid-1960's strategic exchange model Code 50, according to Hoeber, p. 159.

205. See E.S. Quade, in Quade and Boucher, pp. 333-343.

206. Saaty, pp. 415 and following.

207. For example, David Evans's column, *Chicago Tribune*, August 18, 1989, p. 23.

208. Cox, p. 7.

209. Saaty, p. 135.

210. Stockfisch, p. 41.

211. As derived in Bunn and Tsipis.

212. Ermath, in Thibault, p. 606.

213. Hughes, *Military Modeling*, p. 16.

214. Modelski and Thompson, p. 88.

215. Including Clayton J. Thomas, in *Military Modeling* (Hughes, ed.),p. 63.

216. See Owen Gingerich, "The Galileo Affair," *Scientific American*, August 1982.

217. Cooper, Bhat, and LeBlanc, pp. 207-210.

218. Lalley, section 1 (p. 441), provides a strong stylistic contrast to the wartime operations researcher's accounts of search and screening.

219. A sentiment mentioned by Blaug, p. 128.

220. Stockfisch, p. 41 as cited above. Stockfisch first refers to this expression as a "model" on p. 42.

221. Newman, pp. 2138 and 2160.

222. Wohlstetter, p. 220.

223. Thomas, in *Military Modeling* (Hughes, ed.), p. 86.

224. Hoeber, p. 105 note.

225. Koopman, 1980, p. 12.

226. Eliot, *The Love Song of J. Alfred Prufrock*.

227. "The purpose of computing is insight, not numbers."—Hamming, p. 3.

228. Dr. Paul Krueger at FEMA's modeling conference, January 4, 1988.

229. Blaug, p. 248, alluding to Lakatos as indirectly quoted on p. 39.

230. Stockfisch credits this term to Evans, Wallace, and Sutherland. Stockfisch, p. 25.

231. Waddington, p. xiii.

232. Morse, cited in Hughes, *Military Modeling*, p. 68.

233. Builder, p. 109, as well as many of the author's own experiences.

234. Koopman, 1980, p. 134, as well as many of the author's own experiences.

235. Sternhell and Thorndike, p. ix.

236. York, *Race to Oblivion*, p. 211.

237. "Fighting the Drug War: From Arizona to Fairfax," *The Washington Post*, November 6, 1988, p. Al, etc.

238. Exceptions include Richard Feynman's contribution to the shuttle Challenger investigation, reproduced in his book *What Do You Care What Other People Think?*, pp. 220-237, and "Will the Space Telescope Compute?" by M. Mitchell Waldrop in *Science*, 17 March 1989.

239. A comparatively mild example is found in Latterly, (Quade and Boucher, pp. 59-60).

240. Joshua Epstein, *Measuring Military Power: The Case of Soviet Frontal Aviation and The Calculus of Conventional War*.

241. Mahan, p. 8.

242. Brodie, *A Guide to Naval Strategy*, p. 10.

243. Harry G. Summers, Jr., "A Bankrupt Military Strategy," *The Atlantic Monthly*, June 1989, p. 34.

244. Hughes, *Fleet Tactics*, p. 143.
245. Hughes, *Fleet Tactics*, p. 157.
246. Hughes, *Fleet Tactics*, p. 137.
247. A question first propounded by the comedian Bill Cosby.
248. OEG Study 533, p. 12.

Bibliography

Alden, John D. *U.S. Submarine Attacks During World War II.* Annapolis: United States Naval Institute Press, 1989.

Augustine, Norman. *Augustine's Laws.* New York: Viking Penguin, 1986.

Bart, R., and Cohan, L. S. *A Model of Anti-Convoy Effectiveness.* Center for Naval Analyses, Alexandria, Va., 1969.

Beesly, Patrick. *Very Special Intelligence.* New York: Ballantine Books, 1981.

Berkowitz, Raymond S. *Modern Radar.* New York: John Wiley & Sons, 1965.

Berlinski, David. *Systems Analysis: An Essay Concerning the Limitations of Some Mathematical Methods in the Social, Political, and Biological Sciences.* Cambridge, Mass.: MIT Press, 1976.

Blair, Bruce. *Strategic Command and Control.* Washington, D.C.: The Brookings Institution, 1985.

Blaug, Mark. *The Methodology of Economics.* New York: Cambridge University Press, 1980.

Blinder, Alan S. *Economic Policy and the Great Stagflation.* New York: Academic Press, 1979.

Bowerman, Bruce L., and O'Connell, Richard T. *Forecasting and Time Series.* North Scituate, Mass.: Duxbury Press, 1979.

Brodie, Bernard. *A Guide to Naval Strategy*, 5th ed. New York: Frederick A. Praeger, 1965.

Brodie, Bernard. *Strategy in the Missile Age.* Princeton, N.J.: Princeton University Press, 1965.

Brodie, Bernard. *War and Politics.* New York: Macmillan, 1973.

Brookner, Eli. *Radar Technology.* Dedham, Mass.: Artech House, 1977.

Builder, Carl. *The Masks of War.* Baltimore: Johns Hopkins, 1989.

Bunn, Matthew, and Tsipis, Kosta. "Ballistic Missile Guidance and Technical Uncertainties of Countersilo Attacks." Program in Science and Technology for International Security, Department of Physics, MIT, Cambridge, Mass., 1983.

Chatfield, Christopher. *The Analysis of Time Series*, 3d ed. New York: Chapman and Hall, 1984.

Churchill, Winston S. *The Second World War.* Vol. II, *Their Finest Hour.* Boston: Houghton Mifflin, 1949.

Cooper, Leon, Bhat, U. Narayan, and LeBlanc, Larry J. *Introduction to Operations Research Models*. Philadelphia: Saunders, 1977.

Cover, Thomas M., and Keilers, Carroll W. "An Offensive Earned Run Average for Baseball." Operations Research, vol. 25, no. 5., 1977.

Cox, Albert W. *Sonar and Underwater Sound*. Lexington, Mass.: Lexington Books, 1974.

Davis, Lynn Etheridge, and Schilling, Warner R. "All You Ever Wanted to Know About MIRV and ICBM Calculations But Were Not Cleared to Ask." *Journal of Conflict Resolution*, vol. 12, no. 2, June 1973.

Din, Allan. *Arms and Artificial Intelligence*. New York: OxfordUniversity Press, 1987.

Dönitz, Karl. *Memoirs*. Translated by R.H. Stevens. London: Weiden-feld and Nicolson, 1959.

Dönitz, Karl. *The War Diary of the Commander of Submarines*. United States Naval Archives. Microfilm.

Earle, Edward Meade. *Makers of Modern Strategy*. Princeton, N.J.: Princeton University Press, 1943.

Enthoven, Alain, and Smith, K. Wayne. *How Much Is Enough?* NewYork: Harper and Row, 1971.

Epstein, Joshua M. *The Calculus of Conventional War*. Washington, D.C.: The Brookings Institution, 1985.

Epstein, Joshua M. *Measuring Military Power: The Soviet Air Threat to Europe*. Princeton, N.J.: Princeton University Press, 1984.

Feynman, Richard. *What Do You Care What Other People Think?* New York: W.W. Norton and Co., 1988.

Fischer, David Hackett. *Historians' Fallacies*. New York: Harper and Row, 1970.

Fitts, Richard E., *The Strategy of Electromagnetic Conflict*, Los Altos,Peninsula Publishing, 1980.

Galland, Adolf. *The First and the Last.* New York: Ballantine Books, 1957.

Gelb, Arthur, ed. *Applied Optimal Estimation.* Cambridge, Mass.: MIT Press, 1974.

Glass, Leon, and Mackey, Michael C. *From Clocks to Chaos*. Princeton, N.J.: Princeton University Press, 1988.

Gleick, James. *Chaos*. New York: Viking Penguin, 1987.

Gordon, Don E. *Electronic Warfare: Element of Strategy and Multiplier of Combat Power*. New York: Pergamon, 1981.

Gray, Colin S., and Barlow, Jeffrey G. "Inexcusable Restraint: The Decline of American Military Power in the 1970s." *International Security*, vol. 10, no. 2, fall 1985.

Guderian, Heinz. *Panzer Leader*. New York: Ballantine Books, 1957.

Guerlac, Henry E. *Radar in World War II*. American Institute of Physics, 1987.

Hamming, Richard Wesley. *Numerical Methods for Scientists and Engineers*. 2d ed. Mineola, N.Y.: Dover Publications, 1986.

Hinsley, F.H. *British Intelligence in the Second World War*. (Three volumes) London: Her Majesty's Stationery Office and New York: Cambridge University Press, 1984.

Hoeber, Francis P. *Military Applications of Modelling: Selected Case Studies*. New York: Gordon and Breach Science Publishers, 1981.

Hughes, Wayne P., Jr. *Fleet Tactics: Theory and Practice*. Annapolis: Naval Institute Press, 1986.

Hughes, Wayne P., Jr. *Military Modeling*. The Military Operations Research Society, 1984.

Isikoff, Michael. "Fighting the Drug War: From Arizona to Fairfax." *Washington Post*, November 6, 1988.

Jones, R. V. *Most Secret War*. London: Coronet Books, 1979.

Kahn, Herman. *On Thermonuclear War*. Princeton, N.J.: Princeton University Press, 1969.

Kaplan, Abraham. *The Conduct of Inquiry*. New York: Harper and Row, 1963.

Knuth, Donald E. *Seminumerical Algorithms* (volume II of *The Art of Computer Programming*). Reading, Mass.: Addison-Wesley Publishing Company, 1969.

Koopman, Bernard Osgood. "Intuition in Mathematical Operations Research." *Operations Research*, vol. 25, no. 2, 1977.

Koopman, Bernard Osgood. *Search and Screening*. Operations Evaluation Report #56, Operations Evaluation Group, Office of the Chief of Naval Operations, Navy Department, 1946. Rev. ed. New York: Pergamon Press, 1980. References are to the revised edition unless otherwise noted.

Lalley, S.P. "A One-Dimensional Infiltration Game." *Naval Research Logistics*, vol. 35, 1988, pp. 441-446.

Lanchester, Frederick William. *Aircraft in Warfare*. New York: D. Appleton and Company, 1916.

Mac Lane, Saunders. Letters to *Science*, vol. 241, 2 September 1988, and vol. 242, 23 December 1988.

Mahan, Alfred Thayer, *The Influence of Seapower Upon History, 1660-1783*, 2nd edition. Boston: Little, Brown, and Company, 1891.

Meyer, Stephen M. "Soviet National Security Decisionmaking: What Do We Know and What Do We Understand?" collected in Valenta and Potter, *Soviet Decisionmaking for National Security*. London: George Allen and Unwin, 1984.

Mihori, James C., Jr. *U.S.-Japan Security Policy: Protecting the Sea Lines of Communication*. Bachelor's thesis, Massachusetts Institute of Technology, 1984.

Modelski, George, and Thompson, William R. *Seapower in Global Politics, 1494-1993*. Seattle: University of Washington Press, 1988.

Morison, Samuel Eliot. *History of United States Naval Operations in World War II*. Vols. I and X. Boston: Little, Brown, 1956.

Morse, Philip. "ORSA Twenty-Five Years Later." *Operations Research*, vol. 25, no. 2, 1977.

Morse, Philip, and Kimball, George. *Methods of Operations Research*. 1st rev. ed. New York: John Wiley & Sons, 1951.

Murphy, Paul, ed. *Naval Power in Soviet Policy*. Studies in Communist Affairs, vol. 2. Washington, D.C.: United States Government Printing Office, 1978.

National Security Agency. *Battle of the Atlantic*. SRH-008, [Special Research History], Fort Meade, 1977.

Newman, James Roy, ed. *The World of Mathematics*. New York: Simon and Schuster, 1956.

Nitze, Paul, and Sullivan, Leonard. *Securing The Seas: The Soviet Naval Challenge and Western Alliance Options*. Boulder, Colo.: Westview Press, 1979.

Operations Evaluation Group. Study 533, *Effects on U-boat Performance of Intelligence from Decryption of Allied Communication*. Office of the Chief of Naval Operations, 1954.

Pankin, Mark D. "Evaluating Offensive Performance in Baseball." *Operations Research*, vol. 26, no. 4, 1978.

Pierce, J.R. *An Introduction to Information Theory: Symbols, Signals, and Noise*. Mineola, N.Y.: Dover Publications, 1980.

Price, Alfred. *Aircraft versus Submarine*. New York: Jane's, 1980.

Quade, E.S., and Boucher, W.I., eds. *Systems Analysis and Policy Planning*. New York: Elsevier, 1968.

Raisbeck, Gordon. *Information Theory*. Cambridge: MIT Press, 1963. Rehm, Alan. *An Assessment of Military Operations Research in the USSR*. Alexandria, Va.: Professional Paper No. 116, Center for Naval Analyses,1973.

Roskill, Stephen Wentworth. *The War at Sea 1939-1945*. London: Her Majesty's Stationery Office, 1954.

Rössler, Eberhard. *The U-Boat: The Evolution and Technical History of German Submarines*. Annapolis: Naval Institute Press, 1989.

Saaty, Thomas L. *Mathematical Methods of Operations Research*. Mineola, N.Y.: Dover Publications, 1988.

Saaty, Thomas L. *Queueing Theory*. Mineola, N.Y.: Dover Publications, 1983.

Stern, Robert C. *U-Boats in Action*. Carrollton, Tex.: Squadron/Signal Publications, 1977.

Sternhell, Charles M., and Thorndike, Alan M. *Antisubmarine Warfare in World War II*. Operations Evaluation Report #51, Operations Evaluation Group, Office of the Chief of Naval Operations, Navy Department, 1946. Reprinted July 1977 by the Center for Naval Analyses, Alexandria, Va.

Stockfisch, J.A. *Models, Data and War: A Critique of the Study of Conventional Forces*. Santa Monica: R-1526-PR, Rand Corporation 1975.

Summers, Harry G. "A Bankrupt Military Strategy." *The Atlantic Monthly*, June 1989.

Terraine, John. *The U-Boat Wars, 1916-1945*. New York: Putnam, 1989.

Thibault, George Edward, ed. The Art and Practice of Military Strategy. Washington, D.C.: National Defense University, 1984.

Tidman, Keith R. *The Operations Evaluation Group*. Annapolis: Naval Institute Press, 1984.

Ullman, John E. *Quantitative Methods in Management*. New York: McGraw-Hill, 1976.

United States Navy, *Marine Climatic Atlas of the World*, vol. 1, 1955. United States Strategic Bombing Survey, *German Submarine Industry Report*, 1947.

Valenta, Jiri, and Potter, William. *Soviet Decision-making for National Security*. London: George Allen and Unwin, 1984.

Waddington, Conrad Hall. *O.R. in World War 2*. Old Woking, U.K.: Unwin Brothers Limited, 1973.

Waldrop, M. Mitchell. "Will the Space Telescope Compute?" *Science*, March 17, 1989.

Washburn, Alan R. *Search and Detection*. Arlington, Va.: Operations Research Society of America, 1981.

Wohlstetter, Albert J., ed. *Swords from Plowshares*. Chicago: University of Chicago Press, 1979.

York, Herbert F. *Race to Oblivion*. New York: Simon and Schuster, 1970.

York, Herbert F. "The Debate over the Hydrogen Bomb," in Herbert F. York (comp.) *Progress in Arms Control?*, San Francisco: W.H. Freeman, 1979.

Index

A

Airborne radar. *See* Radar

Aircraft

Allied use of, 14, 16, 34, 37, 140

attrition of, 67, 84

ferret planes, 34

probability of U-boat detection by, 80

saddle point in search tactics of, 84

search efficiency of, 10, 20-21

See also Air fleet; Balanced-search principle; Balloon strategy; bombing, strategic; Search-width concept

Air fleet

Bay Patrol by, 16, 29, 34, 46, 148

Bay Patrol losses, 84

as convoy escorts, 21, 65

coordination efficiency for, 19-21

decision for daytime versus nighttime search by, 81-86

effect of Allied, 159-160

effect of decryption for, 46

offensive search by, 29, 64-67

search efficiency of, 91-96

tactics of, 35

See also Balanced force concept; Clean sweep search; Confetti search; U-boat campaign

Alden, John D., 137n

Alien logic theory, 165

Antenna, directional, 31, 37

Anti-aircraft guns, 16, 34, 38, 84

on merchant ships, 105

See also FLAK (Flieger Abwehr Kannonen) guns

Antisubmarine warfare (ASW), 10, 32, 39, 165, 172, 175

Aphrodite decoy, 34

ASG (American S-band) radar, 71

ASV (anti-surface vessel) radar

ASV Mark I radar, 29, 30

ASV Mark II radar, 16, 30, 96

detection by Metox of, 32-33, 73, 98, 100

L-band version, 52-53, 55, 59, 71

See also Rotterdam Gerät

ASV Mark III radar, 16, 35-36, 73

range of, 71

S-band version, 33-34, 55, 59, 65-66, 96, 151

sweep width of, 98-99, 102

U-boat response to, 146

See also Radar wavelength

ASW. *See* Antisubmarine warfare (ASW)

The Author

Brian McCue researched and wrote this book while a Senior Fellow of the Strategic Capabilities Assessment Center, Institute for National Strategic Studies, National Defense University, at Fort McNair in Washington, DC. Dr. McCue was also a systems analyst with McDonnell Douglas, and earlier with SRA Corporation, both in Arlington, Virginia. His doctorate and SM in Political Science are from the Massachusetts Institute of Technology. He also earned a BA in Mathematics at Hamilton College.

U-Boats in the Bay of Biscay:
An essay in Operations Analysis

Cover art by Alphachimp Studio, Inc.
Type IX U-Boat photo courtesy of *www.uboatarchive.net* and
Captain Jerry Mason, USN (ret.)
Graphics by Karen Bostick, Editorial Experts, Inc.
Additional Graphics by Alidade Press

Special Credits
NDU Press Editor: George C. Maerz
Alidade Press Editor: James F. Miskel
Manuscript Editor: Mary Stoughton, Editorial Experts, Inc.
Alidade Press Manuscript Editor: Jeffrey R. Cares
Indexer: Shirley Kessel, Primary Sources Research

A special thanks to Mr. Will O'Neil

Editorial Readers:
Ari W. Epstein, Massachusetts Institute of Technology
Dr. Theodore A. Postol, Massachusetts Institute of Technology
Dr. Alan R. Washburn, Naval Postgraduate School

All royalties from this book are donated to the
Navy and Marine Corps Relief Society
www.nmcrs.org

www.ingramcontent.com/pod-product-compliance
Lightning Source LLC
Chambersburg PA
CBHW031944170526
45157CB00002B/378